Books Hospitality

Chefs And Summer Balls

With a slice of Turkey

Copyright © 2021 on behalf of Hospitality action

All rights reserved. No part of this publication may be reproduced, distributed, or transmitted in any form or by any means, including photocopying, recording, or other electronic or mechanical methods, without the prior written permission of the publisher, except in the case of brief quotations embodied in critical reviews and certain other non-commercial uses permitted by copyright law. For permission requests, write to the publisher.

Acknowledgments

A massive thank you to:
Mark Lewis, CEO Hospitality Action;
Michel Roux Jr;
Kenny Child, Hospitality Action Ambassador, who has now extinguished his keyboard, as the smoke pouring from his laptop was making his eyes sting in producing great stories for 'Chefs and Summer Balls, with a slice of turkey;
Anna Ilaria Crysel, for editing and polishing the manuscript;
Anonymous, frank and honest. #wevegotyou
Catrina Pengelley MIH, lovely reflection on how hotels are beautiful;
Tom Burton, Director of Apprenticeships, NOCN.
Tony Oxley, Apprenticeship Trainer at Milton Keynes College;
Gary Maclean, City of Glasgow College Food Ambassador, Author, MasterChef the Professionals winner 2016;
Steve Thorpe; Hospitality Catering and Educational Consultant;
John Holden FMBII BA Hons, food and beverage Lecturer at Tameside College;
Conor Perrott, studying a Bachelor of Arts, at the University of Derby;

Raginee Scudamore, Chef at University of Buckingham (BAME Apprenticeship winner 2020);
Henal, great dancer, and Reena Chotai, choreographer Hospitality Action Ambassadors, Owners/Operators at Red Cup Café;
Charlie Hodson, Owner Hodson & Co Cheese Rooms;
And to all the fabulous contributors to this amazing book!

Love,
Steve West

Index

1. Foreword by Mark Lewis
3. Foreword by Michel Roux Jr
4. Introduction
16 Great work so far
22 Love Hospitality, but why? Fallen out of love by Steve West
33 Simply Depression, the Truth Behind the Eyes by Anonymous
39 Blessed by Kenny Child
48 My Silent Journey by Charlie Hodson
52 Tales from the Pass by Gary Maclean
58 Inspiration by Kenny Child
65 Nerdy Hotelier by Catrina Pengelley
70 Hospitality Rocks by John Holden
73 Never Give Up by Raginee Scudamore
80 Tales of SAS Detachment 1983, "Brazilian Salami" by Kenny Child

89	Never Lie to Your Chef by Kenny Child
92	Misplaced Loyalty by Kenny Child
96	Oops the Prank Was Expensive! by Kenny Child
98	Are You Sure That's Cooked Crofty? by Kenny Child
100	One Summer in London, a Chef's Olympic Tale by Tom Burton
119	The Humanity of Hospitality by Conor Perrott
134	A Trip to the Culinary Olympics by Steve Thorpe
142	Nightmare on Chef Street! by Kenny Child
148	Ghostly Goings On by Kenny Child
151	Baptism by Fire by John Sullivan
153	Them's the Rules by Tony Oxley
160	Head Hunted by Kenny Child
162	Chocolate Leaves, Has that Young Man Lost It? by Kenny Child
167	Salmon on the Parade Square by Kenny Child

168 Be Careful, Chef, That Knife is Sharp by Kenny Child

173 The Colourful Linguistic Challenges Be Careful What You Wish For, Tendewberrymud by Steve West

186 Mealtime by John Sullivan

189 Don't Take the Pee by Kenny Child

193 Our Story: How Not to Kill Your Partner Working With Your Other Half by Reena and Henal Chotai

200 End Ex End Ex by Kenny Child

205 Getting My Family Barred From a Restaurant... The Walk of Shame by Steve West

213 The Cook, Soldier Moth, EHP and the Groin by Steve West

223 Kitchen, Service, and Shouting Orders for Dummies by Steve West

242 Hierarchy: a Lesson for the New Apprentice by Steve West

261 Necessity is the Mother of Intervention of Invention by Steve West

272 Recruiting Staff as Freud Enters the Picture by Steve West

304 Negotiating the Kitchen Without Killing Oneself by Steve West

316 Computers Are the New World, or Pain in the... by Steve West

323 Pigs Ear, Sugary Silk Peach by Steve West

340 The Housekeeper Shouts, is it a Water Bottle, is it a Rugby Ball? Oh My God, it's Neither! by Steve West

347 Give Us a Lift by Steve West

357 The Grosvenor Hovel by Barrie West

376 The Story Never End

Foreword
Mark Lewis

In the darkest times of the past year, when hospitality people have been at their lowest ebb, it's often been the camaraderie they've enjoyed with friends that has got them through another day.

Deprived of the chance to meet up in person for a pint or a bite with pals, many have turned to social media for friendship and reassurance. My social media channel of choice is Twitter, which I've seen morph into a mutual support mechanism for hospitality workers struggling with the existential threat Covid poses to careers and businesses. New friendships have been forged, old ones reinforced, and humour, banter and stories swapped.

Steve West had the inspirational idea of capturing these stories and turning them into a fundraising opportunity for Hospitality Action. More power to him for embarking on this project – and thank you to all the chefs, restaurateurs and hoteliers who've spared precious time to recount to him their favourite industry stories.

Foreword
Michel Roux Jr

I've been quoted, accurately, as saying that the past year has been my worst ever. Having to deal with the nightmare of the Covid-19 pandemic, it's been far from a picnic. But I do believe in the resilience of the industry and the dedication of my team to deliver a fabulous experience for our guests as they return to eat and drink.

With the passing of both my uncle and father in the past year too, I've had the time to think about everything I, and hospitality in general, owes them, and my cousin Alain and I are proud to build on the legacy of two of the world's greatest chefs through our restaurants and the Roux Scholarship.

It can only get better!

Introduction
Steve West

Mark and Michel, thank you.

The idea of this book and the value it brings, strongly emphasises the strong bond we share with hospitality people.

Twitter is full of words, opinions and retorts which are linked by threads, images, stories, and it's a tool used for many differing personal reasons.

The concept of the book came about because it seemed a crying shame these words were wasted in the constant stream and flow of comments, likes, and follows. It became apparent these thoughts were something which could be used as a positive, a tool for good, for helping a brother or sister out.

It worked, the words became more powerful and stronger as connections developed, the positivity started to bloom into something more than anyone could dream of.

These pages were once blank, white with no words, but as life has an amazing way of becoming better, pages started to fill. The wonder of powerful words is not to be taken for granted.

Chefs and Summer Balls is a collection of stories for those familiar with the wonderful world of Hospitality.

Indeed, even if you are unfamiliar, you will be able to read through it and discover a world not seen, that can only be lived, to understand where hospitality people come from, why they do it, and what drives them.

Driven by passion, the funds raised from *Chefs and Summer Balls* will go toward helping and supporting our brothers and sisters from the hospitality world who are in hardship and in desperate need of help and support. Hopefully our message will be clear: we have their backs.

Hospitality:

Welcome, generosity, kindness, warmth, friendliness.

Hospitality:

Noun: The friendly and generous reception and entertainment of guests, visitors, or strangers.

From the Oxford English dictionary.

Other academic words include cordial, generous, receptive, kindness, consideration, warmth and inviting.

Frankly, these words are the same I would use to describe my grandmother.

Hospitality is an emotive industry matched by none other, the industry spans hundreds of years, it is steeped in history, and the people that make it give their all under any circumstance: restaurants, events, clubs, pubs, festivals and many more.

They give more than they receive, beyond monetary gain, it is intrinsic to themselves to bring the utmost quality of customer care, and five value to the lives of people they have never met.

The book is powered and written by hospitality for hospitality, and is now going down as a legacy for the future. History will look back and highlight these times in which we live, in a desperate time. Hopefully we will soon be able to say we *lived* them, in a past tense.

Chefs and Summer Balls is an example of many coming together in the compilation of this journal, trying to bring across what the industry means to so many: it is not only deep reflections, but real stories to entertain too.

Behind the heartfelt words are generous and wonderful people giving their time to tell stories to help others.

Folks who work in the industry have an inherent quality that far outweigh the words above. The hospitality world is full of invisible people, unseen, going in each day with fire and brimstone, the silent ones behind the scenes for one aim: to provide a lovely customer experience.

During Saturday nights, or mad Sunday lunchtimes on a bright hot summer day, against all odds, these people do everything in their power to give the best service that customers can buy.

This book tells the story of everyone involved in this beautiful industry: from the hotplate warriors on the pass expediting food, to the waiting staff managed by the supervisor, to the night receptionist or porter consolidating the nights taking bookings and waiting for the phone call for a sneaky club sandwich asked for by the brief tenant in their room with clean sheets and fluffy towels.

Our industry is one that people look forward to on their "Friyays" at the end of the week, when we have just started, and our work has just begun.

The machine never ever stops, 24 hours, whilst the rest sleep: bakers, producers, suppliers, butchers, fishmongers, all work towards the next day's delights for the food, linen and facilities of resources for one reason: the customer.

Our customers are missing us, but hospitality folk are having their lives and passions stolen, taken away to leave us in a state of flux, silent confusion, anxiety, depression on occasions, including plunging families into financial hardship and mental despair.

The rug has been pulled away from under the intrinsic value we bring to other people to give them joy and happiness.

We have been denied our buzz, our services, our reception smiles, the noisy echoing laughter in our pubs and the banter with the locals, and it leaves a massive aching hole.

We need a little time and space, even though it is massively difficult in the boxes we live in with close ones and family.

There is a danger our lives at home could be a tinderbox of arguments and friction, but be we are also learning to grow and learn how to behave with the ones we married or got together with, for better or worse.

There is no one to blame for this, there is no one to point fingers at, but like we have always done as a hospitality family, we need to dig deep, talk to each other to ease burdens no matter how much it hurts, even if we feel helpless on occasion.

These true stories we are bringing to you are not only the testimony of the lives of real hospitality people, but also of their feelings, and thoughts.

In the following pages there are some of the messages from recipients of grants, a testament to not only how Hospitality Action stepped up, but also a journal, a diary of sorts, to highlight humanity in times of deep, profound hardship.

The collection holds dear the human roller coaster ride element of pain, tears, laughter, and the purpose in which we hold precious in our hearts.

Gathering momentum toward the end of 2020 was a massive drive toward raising funds for this great charity.

Many high-profile chefs spoke out about how the hospitality industry was hit first and hardest. The people who gave most in the name of giving and receiving, haven't been rewarded for making customer's lives special, whether it be in a restaurant, or in hotels, where the housekeeping department worked hard to keep the linen fresh the chambermaids industriously ensuring the beds are immaculate, the coffee and tea set up in the room.

The receptionists, soldier-like in their uniforms, with the smile to shame angels, greet the customers, not bringing into work if they have had an argument with their partner that day, or if they've just recovered from the flu.

The indelible smile is for the customer.

Invisible workforce, 3 million plus, and as Marcus Waring said, 'it touches us all', and Michel Roux Jr saying 'it is an eco-system'. There are links and connections not only on a localised community, but to the global community as well.

It shocked us all when in 2020 the numbers in the hospitality family dropped, furlough saved some, but many others slipped through the net, floundering around.

Hospitality Action is that water, saving people in times of distress and hardship.

Small businesses are now struggling to survive, save their staff and their restaurants up and down the country.

Overdramatic? Overstated? Not one bit. The 'work so far' chapter is an example how Hospitality Action has saved people in many ways: saved people on the brink of starving, with bills to pay, has given grants and money to help the members who were desperately down on their luck, they've given money to go back to their hometown.

Even the cheeky Fred Sirieix stood up to be counted as part of the army to raise funds and highlight awareness of this plight.

'Eennveeessible Cheeps, mmm. Even Better than Gordons' he would cry out, in a great attempt, alongside his 'Ooh LA LA sauce to sell @invisiblechips to raise donations: the company started to reach out with many other initiatives to get funds to help others in need. Restaurants opened up and customers were encouraged to buy an extra portion of these wonderful Invisible chips, and of course during lockdown, they could be droned in for a little extra. Can't be too careful eh? But how were they cooked?

Heston Blumenthal and Tom Kerridge showed us how.

Phew, at least we know the secret now. Publess quizzes, auctions, walks for calm, beer and wines donated a portion of money raised, the donations came in thick and fast, but as we all know, money runs out, funds drop, and this cannot happen. This book, will not go out of date, it won't perish, it will live on, not only as a testament of true grit of the hospitality industry, but for the reaching out to those in need, for connecting more than we ever have done before.

If there is any doubt that hospitality is not a family, think on this. Hotels, restaurants date way back, full of history, we are connected like through a big hospitality family tree. Today, we are the grandchildren and great grandchildren borne from deep desire of a lifestyle which is unique to hospitality.

The content of this book is not only just stories, but messages of hope, fuelled by the desire to push on, never give up and to repeat one chapter in here, the Humanity of

Hospitality. It is a roller coaster of that humanity, which is apparent and, against all odds, these are people who still believe and are ready at any given moment to brush themselves down, get back up and open the doors again to deliver humanity for their customers.

Great Work So Far

Kind words from some of our Covid19 Grant recipients:

"Before your email I had a little over £5 left to last until next week, a daunting prospect. Thank you so much for the grant and please keep safe".

***

"As someone who is very proud about what I do, it's extremely difficult not knowing when I'll get back to work. A lot of us live pay check2pay check despite working hard & long hrs. Thank you".

***

"I read your message when I was queuing at the local supermarket with my son & burst into tears. I'm used to giving all the time, I love helping, looking after my team & people. Thank you".

"Thank you so much for your help I have shed a few tears, these times are tough for everyone & I didn't know how we were going to be eating at the end of the week".

***

"I want to extend my sincere gratitude for this grant. You have no idea how far this will go; I've been self-isolating, and money has been really tight. Thank you and stay safe.***

"Your grant couldn't have come at a better time, you've no idea how much I appreciate this. Thanks so much and I hope we can all come out of this soon".

***

"There were tears over the dinner table when I received your email. This money is such a blessing at this difficult time and means I can pay my upcoming bills".

***

"Thank you so much. I've been heartbroken and stressed and I didn't stop crying for a week when I lost my job. Your help means I can now put food in the fridge".

***

"Thank you so much, after working in this industry for 16 years, blood sweat & tears I've never encountered such terrible times, thank you from the bottom of my heart".

***

"Thank you so, so much. I have just £8 in the bank. This grant will give us the chance to get some food and essentials."

***

"I woke up this morning so stressed about finances & have spent the entire week worrying. Waking up to this e-mail has cleared all the anxiety surrounding my finances. I cannot thank you enough.

If you work/worked in hospitality and you're feeling anxious or distressed in lock down, please give our helpline a call: 0808 802 0282 It's open 24/7 and it's here to help at these troubling times #wevegotyou."

***

"Thank you so much, I'll no longer have sleepless nights worrying about whether I will have any money to live on, thank you from us all.

***

"Thank you SO much, you have no idea what this means to me! This covers my mortgage this month and I've been having sleepless nights. Thank you again, it means the world!"

***

"Thank you so much. The difference this will make to my family is honestly immeasurable. We have really been struggling & this is like a little ray of sunshine at a very stressful time. Once again thank you.

***

"I was made redundant, couldn't get any hospitality job & I found myself in a situation where I can no longer afford to pay for my basic living costs like rent & food. I felt severely depressed as I was getting deeper in debt. Your grant is a welcome relief."

***

"I think one of the positives of this outbreak is that the public will have a far greater respect, empathy and affection for the people who provide great hospitality".

Love hospitality, but why? Fallen out of Love
Steve West

…Prompted by the notion of 'falling out love' for the Hospitality Industry, a conversation which started on twitter, people felt jilted, let down at the time of Covid, heartbroken in fact. This notion made me think a great deal about who and what drives this deeply emotive lifestyle industry.

Behind the eyes of Chefs, Waiters, Chambermaids, Kitchen Porters, receptionists, General Managers leading great teams for one purpose and one purpose alone, to make customers happy.

Having been told my heart is firmly on my sleeve, I proudly wear the heart there because I feel it drives me.

Food and its culture is global, worldwide: different countries have different traditions, food indigenous to their countries and no one country really is the same as the other in this aspect. Restaurants around the world, preparing, cooking and serving great dishes, Confit, Tabbouleh, Pasta, Noodles, Sushi as brief examples as each country produces and sells gorgeous and tantalising foods.

There is one commonality, even though the cultures are individual in their own right, the feeling behind the global industry, languages and diversity differ, but the underlying unifying obvious nature is the feeling of being in hospitality wherever one is.

Love infers emotion and it seems to be that emotion which excites us. One cannot see how we feel, we just do that… we feel it. Saturday night services, Sunday Lunches, that strange table of 3 on Valentines night, the buzz, being in the subjective zone when all the magic happens.

The love and drive of the hospitality community is inexplicable, the feeling of love and desire and a common word is intrinsic, a deep-rooted lifestyle as opposed to it being just a job. The thought that haunts me is: why? Why would one fall out of love? Why is love a strong word to use for a career? Yes, to some it is a job, but to others there is an addictive burn which will not go away. The receptionist smiles and remains calm as the throng of the bus has been spewed out on to the tarmac front, bags from the other end are greeted by the hotel porter kindly offering to place the bags onto a flatbed to convey to the lobby to be distributed to the guests. All for the customers.

Greeting with a calm face, the receptionist has all the keys as the expectant guest is waiting for the towel in the room to be turned into a swan, basket of fruit from the pastry section wrapped and tied with a ribbon and an edible trophy for the wonderful customer.

The hard work behind the scenes is 80% planning 20% customer experience. For a momentary time, each customer is welcomed. The hard work becomes the lifestyle. It is the inherent nature for the wonderful people in the industry to give, to help, to go beyond the norm and step up to please, welcome and comfort customers.

It is the world hospitality people live in as the moment is driven by customers, checks coming through the murmuring noise from the restaurant, doors revolving with waiting staff asking questions to the chef, usually when a service is in full flight and the chef has numerous tasks in hand as the checks are piling up and in a split second of time, everybody wants everything at exactly the same time.

The chefs and waiters are living their own moments in this world.

Tension rises, heated debates in the middle of service as the tempers start to rage because communication has broken down in the baking, food smelling sauna.

The music of the cutlery sings out as the waiting staff drop the irons into the plastic, soap filled buckets shouting as they go, 'CHEF, TABLE 9 AWAY PLEASE'.

The brain of a chef is in full automotive state of being in the zone. The place where the mind shuts off from thinking to doing is the order of the day.

This amazing mind state happens when refined years of experience come into play. This is the mind of a chef in a split second, and it is a beautiful place.

Hands, mind and body work in magical unison as movements are fluid, swift and like speedy pistons grabbing pans, place on stove, light, control, fridge door whooshed open and closed as the Duck breast is dropped gently into the warmish pan, as the sizzling is an indicator to the chef that the journey of the plate has started for this dish.

Seabream scored, skin side down into another pan, release spray hissed beforehand onto the metal pan making sure this piece of gorgeous fish is treated with respect, not sticking today as the fish is placed in with loving care.

Garnishes are in the mind of the other chef, watchful and in tandem with the mains are ready to be reheated, oven bound as the Gratin Dauphinoise is revealing the secret aroma of garlic, the starchy cake infused gently with heat as the oven does its magic.

The chat before about keeping it as a pave, slab and not break ever invades the mind of the chef who prepared the potato dish earlier that day.

It was pressed within an inch of its life as the science on this journey is now king.

Tender stem waiting patiently on the side, waiting for its stunning glow of chlorophyll state within seconds away into the salty boiling swimming pool adorned with the handle of the pasta basket.

Wipe down, next check, clean, spray, wipe again, look up at the check grab by the brightly lit, yellowy tinged pass.

Away!

Duck is prodded, fish held in residual heat under the lights, waiting patiently for its partner the duck breast to join it.

Duck out, rest, board, slice, the chef gets excited as the pink is visually sexy and pleases the senses. Yes, sexy.

Gratin hovers onto the plate with the aid of a fish slice, tender stem dropped, drained, plates are spotted, drizzled, dropped sauce in seconds, garnished and sent.

A momentary sense of pride infuses the chef, but this is short lived as the next few checks are ready to be expedited.

Day in and day out, doors are open, chefs get changed, banter between the waiters and bar staff with the background of the radio blaring out has started another day.

Chef trying the light the stove, again, click, click, click, as the aroma of the kitchen is undeniable and is a promise of the day ahead. Mise en place lists singing out like the culinary opera from the night before, decanted pots clank on the pot wash ready to be cleaned out, just like the hundreds of times before.

Buttocks clenched as the service is looming and walk-ins during the Christmas period destroying one's soul, the soup hasn't been passed yet and there is no room on the stove, the oven rammed with slow roast lamb shanks is ready, but the chefs are not. The Mise En place list is getting bigger

This is only one day. A few hours.

This is the life, what is not to love?

Or indeed simply, what is the feeling when falling out of love?

What is the emotional change which says, enough is enough!

The day after.

The story continues as the answer in this particular instance was articulated in a very emotional fashion.

The thought of the emotional engagement established an argument about our relationship with Hospitality, and those that work within it.

The answer to 'falling out of love' with the industry was this, falling out of love felt similar to feeling it was the end of a failing relationship, they couldn't make it work and the tip and tronc issue felt abusive as well.

It was like falling out of love which hit home, it is emotion, a relationship is special, a highly intense one, deeply charged and electric, and which deep down is the best and worst of us.

The correlation between hospitality and the busy service, times which we drive our adrenaline body to extremes is addictive and when away we have a sense of loss, which, for better or worse, we miss so much.

In the conversation, it was obvious when they went on to say, no one comes in the industry for the money/glory. It is a desire in you to give people your best, it requires high levels of commitment, and self-motivation. The crux of this, Before the pandemic, did we accept the fickle mistress hospitality?

Are we finding new loves and passions? How much of these words are used in a relationship?

Emotive words like desire, miss, love and the falling out of, feel like our worlds are collapsing and touch us profoundly. Unlike a computer, switch off, switch on and everything is brightly shining through the screens, Hospitality cannot be switched off and on again.

It will take time, hard work, but it will happen.

In the meantime, there is still some work for those who need it.

I wish you well.

Dear reader, life is too short, find your passion, be alive.

Each story has a personality in its own right. There is a circle of life as the tales, reflections, stories are from the hospitality family and are a true source of inspiration, love for the industry, heartfelt from apprentices, students, lecturers, hospitality driven individuals who are still living the life most do not understand.

Read on, enjoy the world of our family for Hospitality Action.

Simply Depression.
The Truth Behind the Eyes
Anonymous

I knew this was going to be a terrible day one of those days when its best to stay in bed, as everything is bound to take a turn for the worst. I lay -motionless- the furious rain pummelled the window.

I curled up in my navy-blue blanket, holding on tightly, as if my life were held on by a single tread- slowly- crumbling. I sat alone without love, or anyone to hold in this darkest moment of my endless life.

As I lay, alone, a twisted vison surrounded me, suffocating me until I gasped for air continuously, I struggled to breathe, my muscles began to contract, never relaxing, stuck in an endless loop of being in defence, never letting the walls that surround me down.

Not letting anyone in, forever alone. A sharp pain flushed through me drowning in sorrow sinking further into the dark abyss. My mind raced, horrors flashed before my eyes weeping with pain, blood pouring down my face, straining my pale white skin. The dark abyss surrounded me, draining me slowly.

Displaced. Lost. I sat- unperturbed- a victim of death, yet not dead.

A time of love, hope and happiness felt distant now. My heartbeat slowly slipped away, becoming but only a small blip in the loud surroundings. Silenced. I felt my memories fade away, but I suppose it is only human nature to add and subtract from our memories: to recall some and feel that they should be remembered.

The darkness holded onto me as the light backs away, I felt afraid, defeated, broken. Death's sly fingers edged closer to me, beckoning me to embrace him and let go. Banging on the padded walls that surrounded me- trapping me- I prayed, I begged, I pleaded to be saved, but resistance was futile at that moment.

Angels fled as demons emerged from the darkness that enveloped my mind, smiling mischievously. A sudden weakness surged over me, like a crashing wave, the surrounding noise became muffled and out of range, my hands began to tremble uncontrollable.

I was sand, tossed by the tide; homeless, friendless. Desperately searching for a way out, a light, anything that could save me from my potential damnation, my personal hell. My mind overflowing like a treasure chest. The pain of a fresh bullet piercing supple skin trickled through my body.

I trembled with fear-motionless- I am alone in a dark room the faint sound of rain hitting the window muffled in the background, yet I am drawn back to the endless abyss that surrounds me.

Inevitable numbness flowed over me, dominating my body, a sense of inferiority consumed me, causing me to hold my head down in shame.

Discarded. My faith stripped from every inch of my body, left paralysed. I watched as the reality that filled my life crumbled around me, my rose-tinted view began to gradually weaken, as I fell to the ground- helplessly- as the darkness crawled closer to its prey, wearing down every inch of my broken, scared body.

The urge of inflicting deserved pain, went through my mind, I called out, but no one seemed to care. Memories flood into my mind, of times I was with those I care about, surrounded by people but lonely, dark thoughts filling my mind.

Maybe someday I will find someone who will not make me feel lonely, but until that day I sit alone. I await the moment my saving grace will come to me and save me from this meaningless existence. The devil's sly hands getting ever closer as each moment passes, the taste of freedom lingers on the tip of my tongue as I watch the devil's hands draw closer to my neck.

This futile life never seems to end but only pushes me further into a pit of despair and self-loathing, I sit alone in my blank bedroom wishing I could end all this pain and suffering setting myself free at last, but the torment or my family and friends pulls me back making me regret and hate myself for thinking such horrendous thoughts.

I just wish I could stop feeling so numb all the time and just feel the smallest amount of pain like heartbreak or a simple burn, but this wish will never come true, so I decide to stay in my room and hide myself from the rest of the world and protect them from me.

The walls the surround my mind are draw closer suffocating me, the devils hand wrapped around my neck like a noose extracting all the air left in my lungs as the last breath leaves my wounded and fragile body, I sink to the floor weakly, the sound of the rain becomes silent, but the echoes continue burying into every fragile bone in my body as my eyes slowly shutting, I am finally released from the chains of my existence.

Free. Yet still so alone.

Blessed
Kenny Child

Forty plus years as a chef and I feel truly blessed.

My career in hospitality started back in 1977, in the September I started a two-year course to complete my City and Guilds at Great Yarmouth college, during the holidays I worked at the then Hotel Cavendish, a typical seaside hotel, which catered mainly to coach trip customers, the food was standard, and what it did teach me was to work quickly, especially during service.

In the September of 1979 I joined the Royal Airforce as a cook, as the airforce called me.

I was civqual (Civilian Qualified) which meant that I was a senior aircraftsman after 2 months basic and trade training. I quickly gained promotion, which meant that I was an NCO (non-commissioned officer) at a very young age, which meant in effect that I was a head chef.

During my nine years, I worked in most areas, fine dining to large scale operations including 4 months in a tent in the Falkland Islands, I had the honour of cooking for Queen Elizabeth, and various other royals as well as government officials, including Margaret Thatcher.

The forces family is very tight knit, similar to the hospitality family.

I then worked as a head chef in places such as Southwold and Walberswick for the brewer Adnams, as both head chef and chef manager, also there was a stint in Aldeburgh at the well-known Regatta restaurant and bar.

After I met my wonderful wife Diane, I decided that the long hours I was working wouldn't be good for a long-term relationship, so I went to work at a large independent department store in Norwich and took the role of Executive chef looking after the food operation for 3 public restaurants, plus the staff.

During that time with an excellent team, we increased the turnover from £1 million to over £2.5 million. I moved on after 15 years, working in various roles in contract catering. After a job that didn't work out as planned, I was looking for a change. I was approached to take a position in a garden centre as a kitchen manager, I thought it might give me an opportunity to slow down a bit as our daughter Roxann had just qualified as a doctor and my son Curtis had just completed his second year at university.

Well, I might've been slowing down figuratively, but I wasn't slowing down in practice: during busy times I was often covering 30000 steps a day, but on the plus side I had lost a lot of weight, and work was exactly 1 mile from home, so everything was good.

Then a life changing event occurred.

We had just had the first few days of holiday, thankfully not away, but at home.

We had visited my hometown for a big vintage car show, had our usual fish and chips with no worries at all.

On July the 9th 2019, I got up at around the usual time at 7:30, made a coffee and switched the tv on and tried to log on to the iPad. This was a bit strange as I couldn't log on and I was locked out.

No worries, as I used my phone but was having trouble digesting the info but still, I was not worried.

Diane came down and asked if I wanted another coffee and I sat and drank it as usual.

As I stood up to go and get washed and dressed, I went to say something except I knew that the words that were coming out were nothing like I was trying to say. I quickly sat down and an ambulance was called.

The remarkable paramedics were with us very quickly, within about 15 minute. By the time they arrived my speech was getting better with only odd words that were misplaced. They said it was possible that I'd had a mini stroke and we should go to hospital. I was fully aware of everything that was going on, but I was not worried, however the looks on Diane and Curtis' faces told me otherwise. Incredibly I went up and washed, dressed and walked out to the ambulance.

The paramedics advised that we should go to the West Suffolk hospital as they have an excellent stroke unit. On the way they kept me talking about anything and everything, there were no issues, except for the fact that I felt a bit strange, roughly halfway through the 20-mile journey once again, I was talking nonsense and the paramedic in the back with me simply knocked on the window to the driver and instantly the sirens were sounding, and I could hear the radio message "get the stroke team ready patient inbound approximately 15 minutes".

By the time we reached the hospital I was improving again. I was met by a whole team from the stroke unit, then was assessed and within an hour, had a CT scan and also a scan on my neck.

I was admitted to the stroke ward and was to have an MRI the next day, it was said that I had a mini stroke.

The next day I was seen by occupational therapists and physiotherapists and all was well, although I was still mixing up the occasional word.

Later that day, I had my MRI, and once back on the ward my consultant was waiting with my family, he asked if was ok to walk to the reception area with him to view the scan results. He said that by rights I shouldn't be walking because the scan showed that I'd had a significant stroke, caused by a clot. I'd just been unlucky.

I had been told I could return to work on a staged return after a month. The thing that was a bit frustrating was that everyone was saying I looked so well, and I think some even thought the stroke wasn't as big an issue as they'd thought it would be. I had been getting headaches and my short-term memory wasn't great. However, I returned to work supposedly on a phased return, I was getting overwhelmed with the amount of checks coming in to the kitchen long story short, I ended up in ambulance and back in hospital, but a new MRI scan showed no further damage what changed was the advice from the doctors: they told me they couldn't forbid me to work in the kitchen, but that would be the best course of action for my health.

And I thought *That's it!* With 40 plus years as a chef gone, what could I do?

In terms of my job, my GP wouldn't sign me off to return, but the company wouldn't terminate my employment in case I sued for unfair dismissal as it would come under the disability act.

Just over a year after my stroke, I am now working in a supermarket bakery twenty hours a week which is about my limit, due to continuing memory issues difficulties in retaining new info and tiredness, but I know I'm one of the lucky ones.

So, it has to be said that the support of my family has been immense, and from the very start, Forces veterans, some that I hadn't seen for over 30 years, were in regular contact checking on me, so a massive thank you goes to you all.

Once I had accepted that my chef days were done, I was thinking *what next*? I got in touch with a couple of Armed Forces charities for advice on a possibility of any training to find employment. Sadly, I got no support from some of them, but I won't name them, as it's unfair on the good ones. I was not after money, just advice. I got in touch with Charlie Hodson and asked if he had contact details for Hospitality Action, so I could ask for some advice. To my amazement the next day, Mark Lewis the CEO, messaged me on Twitter asking for my number and arranged to call me.

The very fact that Mark took the time to call and chat and left me in no doubt that Hospitality Action would be there to support me in any way that was needed. Their hashtag #wevegotyou is really true. Ambassador, Charlie Hodson @charlieboychef, has been a constant source of support even through his own health issues. So, you can see, Hospitality Action is very important to me and so many others.

I am so lucky to have three families: my blood family, the Armed Forces family, and the Hospitality family.

I am truly blessed.

My Silent Journey
Charlie Hodson

My journey to the multi storey was a silent one, although in the background my phone was constantly ringing with an annoying persistence.

Leave me alone, let me be, let me find peace, I deserve that at least, don't I? Said my selfish voice of despair. It was not a voice of a well-balanced human being, it was the voice of a broken man. A man who had planned to arrive at a point on a Sunday afternoon when the town centre would be empty with less chance of being seen and having any interruptions.

In my bag next to me on the bus seat was my little Burberry bag. A little reminder of how life in the past had been a little more successful.

Its contents, a pressed pair of tailored sorts, polished loafers and a crisp white polo shirt. I had packed a bag so I knew when I had found my peace, I knew I would be wearing what I felt most comfortable in and at my happiest.

But none of the above was meant to be. As the bus pulled up, I looked at my phone and the police were calling me. As they began talking to me, reassuring me that I wasn't in trouble and that they just wanted to know I was safe. What I didn't realise was that my friends were on the beach looking for me (as I had left a post on twitter misleading them to my intentions).

But that is where this story will end (only because what happened next really does not need to be shared as it's what happens at the end that I hope will change a life or two). I was finally found, a few hours, in the Queen Elizabeth Hospital by two police officers and one of my greatest mates, waiting for mental health team to arrive.

Finding my own peace was not the answer I was searching for. Instead, I began to learn how to live.

Learning to be put my past behind me, learning to live with all that I could not change, was in fact my only chance of ever finding happiness.

As ever, Hospitality Action was and has been by my side since I listened to the advice given and in turn, they listened to the tears in my voice as I slowly poured out my soul.

That's when the magic began, that's when my life changed. Just realising that what had happened to me was not my fault and from that day I started to lay down a new foundation.

Life is still a struggle and it's not the easiest road but I am now also free from the grasp of cancer (although that's for another story). Life is good with the help of My Saving Grace (a close-knit group of friends) who did incredible things for me.

Now I am Chef Patron of that very same hospital in Kings Lynn which I visited in my darkest time, and an

Ambassador of Hospitality Action. In both these roles I am so very proud to be able to give back just a little.

For me this book is a journal of not only funny quirky stories of banter, and hospitality life but also of hope, because when I started feeling hope for my life is when the magic really began.

Tales from the Pass
Gary Maclean

Chefs are a hard working bunch, we work long hours, miss out on life, miss countless Saturday nights out and miss most of the family occasions most people take for granted. But there is another side to working in hospitality that very few people hear about. I am talking about the inside world of a chef, we sometimes find ourselves in the most amazing locations and in front of some of the most interesting and famous people alive.

These types of stories are normally only shared from chef to chef. I wanted to share some of this experience with you, readers.

When starting out as a young chef and working in nice places, I would often speak to many people, as back in the day buffets where a must in the big hotels, and I would often be in the dining room on the carvery. During this time I met loads of the great and the good, from actors to singers and bands. To be honest, my main concern was keeping the buffets looking good and topped up and learning how to carve huge joints of meat, and at 16 and 17 even getting the chance to be in charge of the buffet was a big deal. I remember one quiet midweek evening I was having a chat with a young long haired scruffy American lad, he was really chatty and it was a nice change to have time to talk to a guest. When I went back into the kitchen, everyone was going nuts trying to find out what this young guy was talking about. As turns out, this lad was Jon Bon Jovi and I didn't even know who he was! For the record, he was just a normal guy who wanted his dinner and a bit of chat. During that time I was working in the best hotel in Glasgow, so when anything important was happening in the city, the key people stayed with us. It was truly incredible how in a matter of months I went from being a lad who was hanging around street corners, to cooking for the Prime Minister and some of the most famous people on the planet.

My next little tale involves the King of Pop. I was lucky enough to be asked to be a guest chef and do a dinner at the Four Seasons Hotel in New York back in 2000, and as it happens Michael Jackson was living at the hotel at the time. I was there fundamentally to do a celebratory Scottish dinner, but when the job sheet came in for one of his kids' birthday party's I couldn't give up the opportunity to get involved and help out. During that trip I also got up to the penthouse of the hotel where Michael was actually staying in, and I must say it is by far the best view of New York I have ever seen. As a wee foot note, the Scottish dinner went well and the birthday party was such a success the exact same birthday party food was prepared again the very next day.

My next tale from the past finds me at Number 10 downing Street, when I was asked by Prime Minister

Theresa May to provide a very special Burns Supper. Some jobs are easier than others, and most people don't know this but number 10 has not got any resident chefs, so pulling this together from Glasgow was a real challenge, especially when there were no staff on site to help me get this done, and I had to think who could help. As there was absolutely no budget for this I was having to pull in favours left right and centre, so I went through my wee black book and came up with the perfect team. As I had recently just competed in MasterChef the Professionals, I thought I could have the best ever MasterChef pop up. I drafted in some of the London based chefs I was on the show with, but the true claim to fame on this event is that I did the Hugh Grant dance down the stairs of Number 10, seen on the film Love Actually! In the end, we all got invited to meet the PM, who, as it turns out, is a big fan of MasterChef. She actually knew all about the chefs I had brought in from the show and spent about half an hour talking to us about it.

My next experience I wanted to mention was a very recent trip where I was asked to do an event for the British Embassy in Havana, Cuba, and again, with most events abroad it is vital to have most of the logistical work already done at home, menu agreed, food ordered and sorted. I wrote the menu and the food orders and sent them off to the staff at the embassy, and almost the minute I had sent the email I got one back stating that none of the food I wanted was available: this included basic staples like potatoes and any type of meat of fish. After much thought I decided that I would "wing it" and sort the whole event once I got there. I got to Cuba and totally fell in love with the place, the people where amazing and I just loved the culture. The problem for a chef in Cuba is that there is very little food available. So what I did was spend two days going round Havana with a driver, on the hunt for food, even finding myself in more dangerous areas. As it turned out, I had to buy most of my food from the black market, and finding simple things like potatoes and eggs was hours of work, and there was loads of meeting people in dark alleyways. In the end I managed to get enough food to pull together a great event that was attended by some of the most important people in Havana.

Inspiration
Kenny Child

This is for those who inspired me whilst serving in the Royal Air Force

Way back in 1979, on completing my city and guilds certificate, I decided to join the Royal Air Force. In those days, if you were civilian qualified, it meant that you completed basic training as a Senior Aircraftsman, so following 6 weeks at RAF Swinderby, off I went to the RAF school of catering at Hereford, for just 3 weeks to ensure that I could do what my qualifications stated. This was one week in the training kitchen, and then into a mess kitchen, which in my case, fed around 700 trainees.

At the end of this, I was posted into the real world, where I received a massive slice of good fortune as I was off to RAF Wattisham, near Ipswich.

On arrival I was placed into the officer's mess, which normally consisted of about 40/50 covers for 3 meals a day, with a daily changing menu, all fresh food with regular functions, for 100 plus people for formal dinners and around 400 people for balls and banquets.

After around six months, I was moved to the junior ranks mess, where I initially thought that I'd not been up to the task, as from running my shift I would be a junior cook with three corporals, 1 sergeant and a Warrant Officer. That officer was Dave Walklate.

Two of the corporals, Bill Evans and Ken Owens, are still dear friends to this day. They had told me that the Boss was a decent one, that he didn't do any cooking, but was a great chef and just to keep my head down and work hard, as we normally fed around 600 people each mealtime, and when on exercise this increased to over 2000 each mealtime.

After around 4/5 weeks I was summoned to the office. I thought I was in trouble, however I was offered a seat and a brew. Mr Walklate (that's how warrant officers are addressed) asked how I thought I was doing, to which I replied that I wasn't doing bad.

A wry smile crossed his face, he explained that he'd heard good things about me, and that he would soon be completing my first assessment. The Air Force had an appraisal system that involved number rating scale 1/9 now 1 meant actually you shouldn't be in the air force and 9 meant that you were basically perfect. This was assessed on trade performance and personal i.e., appearance behaviour etc., and included a recommendation for promotion if appropriate.

I was taken aback when I was told that I was to achieve 2 x 8 and a special recommendation for promotion. It was further explained that there was no way that a catering officer would be downgraded, and that there was no chance of being promoted at the age of 19 with less than 1 years' service.

He then explained that for the catering officer to sign off the assessment, all of this would have to be justified.

Mr Walklate told me that the following day there was no Corporals on the day shift as they were both to do routine weapon training, and in fact there was only staff even less experienced than me, so I was to lead the shift. I didn't see either the warrant officer or the sergeant at all, until he appeared to check the servery was ready for dinner. To my surprise, the Catering officer was with him, something that was highly unusual. However, I was confident that there was no issue. Then it happened: "Airman, get in my office now", I was told. However, we didn't get that far, as I was stopped by a tirade of expletives, and as I turned round, with him was the catering officer behind him.

The words still lived with me to this day.

"What's wrong with the Chicken Maryland?" I was asked. I replied that I was happy with it.

So, it was explained that instead of sweetcorn beignets I had simply done mini sweetcorn pancakes and that I was a disappointment to him.

I was crestfallen the catering officer didn't seem to know where to look, I muttered "Sorry sir," and I was met with the reply "just f*** off out of my sight" with that they disappeared. The boys in the kitchen just stood there, nobody said a word and just cracked on working.

Around 20 minutes later the catering officer emerged in the corridor and caught my eye and simply said "good work today, well done".

Towards the end of service one of the lads told me that the boss wanted me in the office, and he looked angry still.

I knocked on his door, entered and was ready for a dressing down.

His face was like thunder, but I noticed two cups of tea and some biscuits and suddenly a big grin appeared and I was offered a chair.

The following words left a mark on me that have stayed and became my mantra.

He explained that as a chef he couldn't have produced the standard of food that we had for that amount of people, it was further explained that all of his time in the air force he had worked in officer mess as he was such a good chef, but mass cooking wasn't his thing.

The lesson was never to accept anything less than up to the very best that you can achieve, and that I had proved that I was worthy of the assessment.

For the next few months, I worked the pastry section, and during that time there was an open day, and my parents attended the air show. I was working but they appeared to see me at the kitchen door, and Mr Walklate saw them and invited them to his office for tea and cake.

After they left, I was told that a new sergeant was arriving in the officers mess.

Alan Couzens, who was a fantastic chef, was to arrive and I was to move back to work with him to learn further.

In early 1983 I was called down to see Mr Walklate, which was was strange as I rarely saw him, again it was the tea and biscuit scenario. I was to travel to the Falkland Islands for a four-month detachment.

And more importantly that, off the record, I was to be promoted and posted on my return.

Being such a young corporal would have different challenges, but that's for another time.

I will be forever thankful for the guidance given to me at an early age and am still in touch some 40 years later.

How I became a "Nerdy Hotelier"
Catrina Pengelley

I expect everyone's journey to finding their passion and career starts a bit different. From eight years old, I already knew that I didn't want to work in hotels. Of course, I am sure you can already tell that didn't work out so well for me.

I grew up with a single, young mum who was a Front Desk Manager at a hotel. The perks were awesome: I would spend quite a lot of time causing chaos in the lobby, restaurant or pool areas and I would get to stay in fancy hotel rooms when she worked the night shift. However, my mum worked long hours, had an inconsistent schedule and was always dealing with, for better words, rude people.

Like most children, I went through many stages of career ideas. Marine Biologist, Astronaut, Dancer, Millionaire, you know, the usual ones. Fast forward 10 years later, me in my white coat, covered head to toe in salad dressing, random sauces and a bit of salt and pepper, working my way up to my now current dream of being a chef.

Still having a huge passion for cooking and desperately missing the rush of being behind the line, but working in a kitchen is a tough gig, especially as I was the only girl on the team at 18 years old. So, to relieve some of the stress, I took a second job waitressing, which allowed me to realise how fun it was to interact with the people on the other side of the window. Over the next few years, I took on multiple roles in both the front and back of house. I wanted to learn the ins and outs of the entire business as now, my new dream, was to be a Chef with my very own restaurant.

This led me to a Hospitality Management Diploma program focused on running hotels and restaurants. I was there for the restaurant knowledge and experience, but I thought, there is no harm in learning about different areas of the industry, right?

Wrong! Or at least, wrong depending on how you look at it. I don't know when it started, but all of a sudden, I was weirdly obsessed with everything that I didn't know about hotels. The things that get you saying, 'wow how cool', the 'huh' factor as I call it, had me hooked!

Having worked in many different role in hotels over the past six years, and the weird excitement that I mentioned before, I can definitely escalate my "obsession" to calling it "nerdiness". I have become the type of person to be in a hotel and see a beautiful marble staircase or old wooden panelling and go 'ohhhhhhh' as I stroke it. I walk past an old, Edwardian hotel and need to touch the 'architecture'. Suddenly, I'm 45-minutes late for wherever I was going and am now lost on the 5th Floor... but honestly, I don't want to leave anyway.

Hotels are beautiful. Like people, each so different, with an individual story to tell. I hope to stay 'nerdy', to continue to appreciate the architecture, design, story and uniqueness of each hotel. As I look back, I tell my eight-year-old self that the long hours, inconsistent schedule and rude people are all just little, menial aspects of my job. Hotels and hospitality, its more than just my job, it's my passion.

Hospitality Rocks
John Holden

Well, hospitality is my passion and my life (apart from my family), but when you put that with teaching hospitality it's like being in job heaven.

I would like to share with you why these two professions, hospitality and teaching, make it all worthwhile.

I had been teaching for 4 years, and a mature student wanted to change career, and he had already been told that he could not do a hospitality course.

I went to speak to him, and we managed to get him on a course anyway. He started it and stayed with us through thick and thin, working as well as doing college and fighting financial difficulties to succeed to make it into hospitality.

He finished his level 3 certificate, went out into the industry and started working his way up the ladder.

Some 5-6 years later, he called the college and asked to speak to me, I thought he was going to ask me if I knew of any jobs going or if there were any students who wanted a job.

But no, he started by saying 5-6 years ago he had no support to change my career, that he met me and I gave me the chance, supported him through all three years and supported me since I left, in so many different ways.

He was phoning to say thank you, thank you for my support, dedication, kindness and that if it was not for me he would not be where he is now or be the person he is now.

Well, I was speechless, what could I say to that? I was in tears, I could not speak for at least 2-3 minutes.

I am proud of him to this day, and it's so nice to be acknowledged as being part of his success.

That's hospitality and education making it all worthwhile!!!

Hospitality rocks!

Education helps it rock!!!

Never Give Up
Raginee Scudamore
Winner of BAME Apprentice of the year 2020

I first started working for The University of Buckingham Catering Department as a front of house food service assistant back in 2016. I had not worked since 2002, when I had my first child. My family has since grown up I felt that I wanted to get back to work and try out something completely different but within an industry that I am passionate about.

I have always had a great love for food, right from my childhood, which was sparked and nurtured by my mother and grandmother who taught me how to cook and bake at home. I particularly loved it when I was shown how to cook recipes from my native homeland of Mauritius, where we grew and used many of our local ingredients.

My job was very varied, including directing our front of house staff during service of some very high-profile events such as the University graduation and private dinner parties held by the Vice Chancellor.

During my first work appraisal, while discussing my role with my manager, I realised how much I had developed and how much more I wanted to learn. I was eager to learn new skills and I had always had the ambition to become a chef. I was offered the opportunity to enrol as a commis chef apprentice at Milton Keynes College.

I remember the exhilaration and excitement of the first day, sat in my car in the car park at college. I felt as I did when I left my family in Mauritius all those years ago to start my new life in Great Britain, anxious, nervous and full of anticipation all at the same time.

It was obvious right from the start that I was so much older than the rest of my group and felt that I was at a huge disadvantage. As soon as I met my tutor, I was made to feel completely at ease, he welcomed me as a mature student, embraced my cultural background and made me feel like I belonged in the team.

My life became a whirlwind. Once a week I would attend college and I had plenty of challenging coursework keeping me up often until midnight, or even later. I also had a family to look after.

The other four days were all spent in the university kitchen.

My first day in the kitchen felt like walking into a Gordon Ramsay program. I wondered why chefs had to swear so much. I felt like I was in the way, so I kept my head down and did what I was told. Everything was loud and at a fast pace. In particular, I remembered that I was asked to empty the chip basket which was extremely heavy, and I could hardly lift it.

This made me realise that you need physical strength for this job.

So many memories come to mind whe thinking about my journey.

I learned the hard way that one should never talk back to a head chef! Preparing the salad bar is a two-hour job, but it would take me four. I still have not lived this down.

One time, while I was cooking eggs, the cartons caught fire and in panic I carried the flaming cardboard through the kitchen and nearly burned the whole place down. My apprenticeship nearly ended that day.

Every day I went home exhausted. However, as the days went by, I got used to the shouting, the swearing, the noise and the physical work.

While being taught how to make a lemon drizzle pudding, I learned that even the head chef can make a mistake when he finished the dish by sprinkling it with salt instead of sugar. He tried to make it right with a bit of scraping and re-sugaring, but incredibly he made the same mistake again. The pudding obviously ended up in the bin!

Over the next 17 months, I worked towards my learning goals, with lots of support from the staff at work, and I had regular meetings with my assessor in the workplace which really helped me to keep on track with my qualification aims.

Part of my learning also involved trips to different hospitality venues such as Borough and Billingsgate Markets and Corinthia and the Ritz Hotels. These trips were exciting and inspirational, and made me realise how much I wanted to be the very best I could be in this industry.

The college visit to the Ritz in London left me wanting to learn more, so I applied for a weeklong work experience there, and three months later I had that unique opportunity to be there. I booked a hotel in London as it was too far to commute daily, and I wanted to focus and reflect on my time there. I am very fortunate to have done this experience and thoroughly enjoyed it.

The Ritz's kitchens are much busier than what I was used to at the university. The pressure to achieve precision and perfection was incredibly high. They used a vast selection of premium quality and exquisite ingredients, some of which I had never come across before, like delicate flowers that were meant for the dishes. The expectation to deliver at a fast pace is crucial.

There is an etiquette at work and a sense of glamour and sophistication in all the presentation of their dishes. There are no distractions and there is always an incredible need to stay focused. There are many more things I took away from my experience, some of them I implemented in my own workplace, such as the need to deep clean one's workstation twice a day to avoid cross contamination, especially now during the Covid-19 pandemic. The strong organisational skills needed, and methodical working is something that I have integrated into my daily routines.

At the end of my apprenticeship, I won the BAME Apprentice Award for Hospitality. Winning this was like a dream, considering that as an older woman I walked into college on the first day surrounded by teenagers and thought I didn't belong there. All the staff worked so hard to help me and make me feel like I fit in.

They never gave up on me, and now I have a wonderful job and wouldn't change anything for the world. This apprenticeship has given me recognition, a title, a good salary and freedom to progress in the hospitality industry.

I have even learned how to swear like a chef!

Tales of SAS Detachment 1983 "Brazilian Salami"
Kenny Child

In mid-1983 I arrived at RAF Stanley following a luxurious 10-day cruise of the SS Uganda and the first night on the floating hotel which was later to be used as a prison, which gives some idea of the accommodation.

I arrived in the portacabin which served as the catering headquarters.

The Squadron leader informed me that as I would be promoted upon my return to the UK, I was to take over and run SAS det. To be clear, the SAS bit had nothing to do with the special forces, it stood for 'soup and sandwiches'.

Our role was to serve lunch daily for those that were working on the airfield: there were no tables or even cutlery, alongside this we would carry out the inflight catering for the Hercules aircraft based at Stanley and the daily air bridge to and from ascension.

Our daily shift would normally start around 5 am, serve lunch and then wait for the flight to arrive at around 4pm and replenish the supplies.

The Menu was:

Homemade soup

Doorstep sandwiches

Pizza

Erm… that's it!

Now to set the scene, the only food item that didn't come from a tin was the bread, this was made at the Army bakery in port Stanley, and I have to say, it was decent.

Even the margarine was tinned.

Anyhow, one bonus was that along with the ration packets we got included chocolate that was also tinned but it gave me bargaining power to obtain other stuff like batteries for our cassette player and warmer clothing than our standard issue, but that's another tale.

The pizza dough was not easy to make bearing in mind we had no olive oils, and the fat was hard to rub in due to the cold temperature and having no mixer.

We would serve around 300 people each lunch time, and the pizza was cooked on large trays and cut into squares. The cheese was also processed and tinned, bearing in mind that also the ovens had no thermostats, just the gas burner underneath.

During my time there was a VIP visit as some government minister was visiting who would not have a clue about what it was really like. I didn't always quite follow the rules, so it seemed a bit risky, but I decided to act outside of the orders I'd had been given.

For this day I had been given some loins of pork and lamb, and the idea was to serve them hot in sandwiches, but the thing was I was to hold them back so that when the VIP party arrived, they would have these splendid fillings for sandwiches.

Anyhow, that last bit of information slipped my mind so as normal the queue was forming just before 11:30 opening time ready for the normal fare of pizza and sandwiches with such fantastic fillings as meat spread, cheese jam etc (no wonder most went for pizza) however the troops were chuffed to see this array of hot freshly cooked joints of meat.

Unfortunately, they had all been consumed before the party arrived. My lads were worried, but I just took matters in my own hands and said I'd take the flak. In they came, the government minister approached and introduced himself and asked about the food we served, as I was chatting to him, I could see the catering officer looking for the fresh meat and starting to fume quietly.

"What's on the Pizza, Chef?"

Now, I don't really know why I said it, but then main topping was a tinned food called bacon grill, which is a bit like spam but with a bacon taste, so I explained it like this.

"I have a contact in the Royal Navy that manages to get me specially sourced Brazilian Salami, and it's the most popular thing we serve".

The senior officers with him both confused and amused, and he said, "how splendid that they get a treat now and again" after trying the pizza. I was amazed that he told me that it was fantastic, and thanked us and that the feedback from those that were in the tent was lovely to hear.

As they were leaving, the Catering officer growled "my office in one hour".

That's it, I thought, *no promotion for me.* So, it was with trepidation that I approached it, and as I went in, I saw my squadron leader and a couple senior officers with him, and my heart sank.

"So, airman explain yourself".

"Sorry sir how do you mean? "I replied.

"What happened to the pork and lamb that we were meant to have?"

"I have to take responsibility, sir, I must have misunderstood what time you were arriving, and we ran out".

"Really? That's unusual for you to cock up. Tell us the truth".

So I went for it.

"Well sir, I take responsibility, but I decided that the troops deserved a treat but hoped that it wouldn't have all gone when you arrived. But unfortunately, it was more popular than I'd imagined so I tried to impress the Government Minister with the salami Pizza".

There I stood in front of three officers waiting for the worst, but I was asked to leave and wait outside in the outer office. As I waited with the admin guys, there was a mixture of laughter but I also had a sense of dread, thinking I was an idiot and my promotion would be gone.

It seemed I waited for ages, but soon I was called back in.

The boss spoke "you went completely against what you were told to do so you should appreciate that you're in trouble, however the minister was raving about his pizza and how wonderful the food our cooks are producing, so we will let it go, however we do have one problem. He wants some salami to take back to the UK !!"

So, whilst I was treated with great respect by a lot of the troops that had heard of the tale, I was warned by the catering officer that he considered me a loose cannon and to watch myself.

As it turned out, I was left alone to wheel and deal as long as there was no financial gain but for the morale of the diners.

Brazilian Salami Pizza Recipe

Heat oven to hot (no thermostat but you can adjust the gas by getting on your knees)

For the dough

4 hand bowls of flour (plain)

1 sachet of dried yeast found in the ration pack.

2 tins of margarine (no olive oil so improvise)

Water from the bowser hopefully it's been filled up.

Rub in the margarine into the flour that will take ages as it will be rock hard and no food mixer.

add the water mixed with yeast, once mixed roll out the dough onto the biggest baking tray that will fit in the oven.

Topping

Tomato ketchup (that's all we had)

Chopped onion.

Tinned compo sausages (as many as we could spare)

Tinned bacon grill (also known as Brazilian salami.

Tinned processed cheese (grated)

Spread over the dough evenly.

Now whether this is thin or deep crust really depends on how many degrees above freezing the tent gets before it's time to be cooked.

Cook until golden brown.

Normally you're thanked with a cheers mukka, if you can't understand them you may assume, they're Raf Regiment.

Never Lie to Your Chef
Kenny Child

To set the scene, in the late 1980s I was the NCO (non-commissioned officer) in charge of the combined mess, which was attached to Raf Coltishall, about ten miles away. My staff lived there and were bussed in for their shifts, I however lived in Great Yarmouth around 40 miles away.

At 3pm the late shift arrived for duty as normal, I wandered into my office and on my desk was a bottle of tablets, so I went into the kitchen and asked who's they were.

Leading aircraftsman Smith pipes up "they're mine chef, I'm not feeling great" he assured me he was ok to work so I ignored him, thinking *that's a big hint*.

Another member of staff a couple of minutes later asked to see me in private.

The upshot was that it was Lac Smith's first wedding anniversary and had wanted to take his wife for dinner, so he thought he was trying it on.

The thing that annoyed me was that if he wanted to change shifts it wouldn't have been a big issue, it could have been sorted but he'd said nothing.

At 5pm, before heading off home I checked on him to ensure that he was ok, and I left. Five minutes after I get home, the phone rings and the guard room tells me that the cook says he's been sick and has diarrhoea, so I tell them that I'm returning to work but the cook has to wait until I get there.

I was fuming as I knew it had been planned. I formed a plan during my 40-minute journey. On arrival I stopped at the guardroom and phoned the medical centre at Coltishall and spoke with them.

I informed them that I was concerned that the cook had food poisoning and wanted him to be admitted for observation and to take stool samples.

I then phoned MT (transport) to collect him and not to take him home, but straight to the medical centre.

When I got to the kitchen he was already changed from his whites and into his uniform and ready to go home.

The upshot was, no meal out with his wife, he was prodded and poked for 48 hours as they knew full well it was a set up.

Bless him if he ever knew that I was aware of his plan if he was, he never mentioned it again. I had to do a double shifts, which was a small price to pay.

Never try to con a guy that's seen every blag going.

(Name changed to protect id)

Misplaced Loyalty
Kenny Child

In 1991, I was head chef in a very popular restaurant and bar in a well-known Suffolk coastal town, and we were known for welcoming numerous celebrities and titled folk on a daily basis, it had been a great success, however the owners had tried to replicate the formula in two other locations. Sadly it wasn't working and they were losing money, but the biggest issue was that they had used the original restaurant as collateral to finance this expansion.

As the company was entering a cva, the final trading day was with us, in the morning our wine supplier arrived to reclaim the stock that obviously wasn't paid for, I spoke to them and suggested that there was no way of proving what had or hadn't been paid for in the cellar.

Reluctantly, they agreed to return on the Monday for a stock count. So that Saturday night service was fantastic, even at a reduced menu, with loads of deals and all payments in cash, of course.

So, it was apparent that the owners couldn't pay what was owed in terms of unpaid salary and holidays, however they suggested that I keep the company car that I was using (a Fiat Panda).

I was quickly employed as a head chef for a newly refurbished hotel a few miles further up the coast, so there were no issues for me brigade and all was good.

Then it happened: the kitchen faced a main road and very early during an evening service, a massive crash was heard. One of the kps shouted to me "chef it's your car!!"

Outside I went, and I saw my car had been hit and spun round. I called the police, and it was clear that the cause of the crash was a drunk driver, so took my details and arranged for the car to be taken to a local garage.

Fast forward two days, I received a call from reception to inform me that the police would like to speak to me.

We went to my office to chat and quickly I was horrified to hear what I was being told: the insurance that I'd provided was not valid, as the premium payments had not been kept up, so I was facing a charge of no insurance. If that wasn't enough, I was then told that in fact the car was on finance and was the subject of a repossession order, which, of course, I was totally unaware of.

The police fortunately completely understood that I was unaware of any of these issues.

Also, the drunk driver was uninsured, so it was decided that losing the car as well as wages, along with dealing with the finance company was a punishment enough, so no action was being taken against me.

Harsh lessons learnt, and surprisingly I still remained friends with the owners, but politely declined an off to work with them on a new venture a couple of years later.

Oops That Prank Was Expensive!
Kenny Child

Way back in 1990 I was sous chef in a fine dining establishment, it was mega busy, but we worked hard and played hard.

A very young commis used to come into the car park in his Golf GTI, with music blaring out. The car was bought by his parents, which for some unknown reason really wound up the head chef, maybe it was because he had a battered old Citroen, but who knows.

One Saturday morning that chef decided to put pay to the music, and on realising that the VW was unlocked, he took his chance: he removed the big speaker from the boot and hid it in the staff accommodation. Lunch service had finished so the head chef, myself and the junior sous left for the pub. Just on to the second pint, the landlord came through and said that the hotel manager thought that we should go back as the police were there with a distraught chef, so we all went back and the police instantly wanted to know if we had seen anything regarding the theft of the speakers.

The head chef immediately confessed that it was him playing a prank and he would go fetch it.

Now, to be totally honest, the head chef, myself and the junior sous had all started at the hotel around the same time and two things were apparent: A, we didn't like the head chef, him and B, we didn't rate him, as he always avoided service and left us to run the pass. So, we just left him to deal with it.

Then came the kickback and the thing that amused us the most: the stereo system wasn't by any means standard, so it wasn't just a case of re fixing the wiring it had to be reinstalled and calibrated by an auto electrician, which resulted in a nice little bill for around £40.

Are You Sure That's Cooked Crofty?
Kenny Child

In 1977 I knew a lad in college, whose name was Steven Croft. We were in a bakery practical lesson and we were making fruit pie, the filling was not important so sugar pastry was made, pie sealed and in to the oven as per the recipe. We were fortunate that we had our own bench and oven, after the amount of time per recipe, some of us were taking the pies out, a couple of us saw Steven taking the pie out of the oven, it obviously wasn't cooked and when we told him so, he replied "Nah, you lot are always on the wind up it's been in 30 minutes".

Imagine our horror when he removed the pastry ring, to see it had collapsed!

Mr Jeremy, our fantastic tutor, came over to see what the fuss was all about,

"So, what do you think happened son?"

"I followed the recipe, sir".

"Did you think about pre heating the oven? You haven't switched it on".

Maybe he wasn't cut out to be a chef. All I know is that he didn't reappear for the second year of the course.

One Summer in London
A Chef's Olympic Tale
Tom Burton

Hull, December 2010.

"Do you fancy working at the Olympics", went the immortal words which would change my life for the next two years. I was part of a pub consultation project, my first foray into working for head office rather than my original day job at Doncaster Racecourse. It wasn't exactly the bright lights of London that started this journey, more of a snowy, very cold, old working men's club on a council estate in Hull.

Over the space of two days (two days where we served no "actual" customers), the discussion kicked around the Olympics and potential of catering at one of the venues.

"We have won the tender for Eaton, we need a project team", "and would you consider working in London?"

These sorts of serious career conversations I have found usually happen over a pint in a hotel bar. The "sell" by the group Catering Director arrived, "it will be really good for your career", "you have just got married, and it will be great on your CV", "it will give you an amazing once in a lifetime experience".

Not sure if it was the hotel lager, or just me saying anything to get out of another pub consultation, but I agreed to go home and have a think about it. I was 27.

March 2011 – Feet first

"Tom, you're going to have to either concentrate on the Olympics or on Doncaster, but not both", Trying to split the early days of Olympic planning and working as a Sous Chef at Doncaster Racecourse was becoming difficult.

Rightly, as many times over the last decade, my Exec Chef's advice was to prove a turning point. The next day I was seconded from my racecourse role and became the Culinary Project Lead for the head office's newly formed Olympic project team.

Since those early conversations in Hull, things had somewhat advanced. Not only were we now planning for the rowing events at Eaton Dorney, but had also won a similar tender for Greenwich Park, so we had to deal with two venues. It seemed a logical concept, if we had to go through an 18-month project, we could just double up on everything and maximise the economies of scale. We also came very close to a third site at horse guards' parade, but a third site may well have finished me off!

"So how exactly are we going to do this?" was a thought which circled around in my mind almost constantly in the spring of 2011.

Thankfully I wasn't alone, the catering director had put together a strong initial project team, myself for the kitchen team, a front of house manager, a catering accountant, a project manager and human resources. All of which started to pull ideas together.

My urgent brief was menus, Olympic compliant menus. "Great," I thought, "easy!" I have been writing menus since I started in the game. "How many do we need?" "About thirty for each venue".

"There are a few things you need to know Tom". The "few things" was the first introduction to how an Olympic organising committee works. Nothing, not a single item or dish could be repeated across any of the eight rolling days of menus. Nothing could be sourced without "Red Tractor", "RSPCA Assurance", "Marine conservation society approval", "local to London" or "ethically sourced Olympic-approved" to name just a fraction of the red tape

I had to navigate, "oh and nothing red which could stain the 17th Century former royal residence". "Say again?"

Summer 2011 – Designing kitchens, writing menus.

30 menus on the face of it seem relatively straight forward when you say it, but exactly where were we supposed to cook the items on these menus? Drafting any initial menus and dishes always starts with just a blank piece of paper, but normally as a chef you know you have this oven, that fridge, this walk-in, those plates. At both Eton and Greenwich, we had a field, oh and that 17th Century royal residence.

If you want to cater for any event in a field, or anywhere that isn't bricks and mortar, you need specialist kitchen providers. We worked with some of the best, experienced in building kitchens in every type of environment, particularly in the cases of this kitchen supplier, war zones.

This became a little chicken and egg, they would say "what equipment do you need Tom?", "what do you have" would be my usual response. CAD drawings, specs, site walks, lighting, power generators and bottled gas all consumed the early days of the menu design.

I wrote and wrote, identifying products, sketching dishes, noting costs, revising costs and worked out nutritional analysis for the athletes on their menus.

After a couple of months, we had initial approval from the organising committee of the games to present our first draft. Approximately 600 items and dishes, breakfast items for athletes, workforce meals, overnight snacks and one of the most important menus sets, the Games Family. The Games Family are the V-VIPs. The Royal family, the celebrities, Kings and queens of Europe, the great and the good of the Olympic games. All to be served in that 17th Century Royal palace.

The menu presentation and tasting test were due on a Thursday and Friday in July. By this time, I had brought together some other of the group's chefs to support as well as some first-class freelance chefs, experienced not only in high end catering, but also in handling the volume of food required.

We decamped to Windsor Racecourse, which was a short hop across the river from the actual Eton site, giving the whole project a sort of reality check. This was the first real big step for the culinary element of the project. Starting on a Saturday afternoon we set up the kitchen, briefed the team of five chefs and set out the prep lists and allocated jobs. We worked away all week and by the afternoon of Wednesday, we were ready for the first testing on Thursday.

The catering, cleaning and waste division of The London Organising Committee of the Olympic Games or LOCOG arrived in a team of three. Appointed from previous games organising committees, and large event venues. The head of the trio had a somewhat headmistress aura and was very much the chief of the three.

The committee began working their way through the menu, scrutinising even the sugar sticks on the drink's stations.

6 hours later, we were dejected and beaten. Our menus and presentation went down as bad as it could have gone. We were way off the mark.

We had applied logic and safety, so we went with secure and low risk. It wasn't what they wanted, they hated it and they were disappointed.

We salvaged, somehow, the Friday test, which was for the workforce and athletes. But the Games Family menus from the first day were pulled apart, criticised and destroyed.

Lingfield Racecourse, 4am 27th September 2011.

We had been given a reprieve. We had missed the mark for the Game Family menus. I had four weeks to re-design, cost out, and finalise the dishes before the cut off at the end of September. I went all in.

I researched and found new suppliers. An artesian patisserie with a royal warrant, a north Yorkshire smoked food producer, organic mozzarella from Kent, Cheesemongers, fishmongers, specialist butcher and a game supplier from one of the royal parks. I suppose the chef equivalent of pushing all your poker chips into the middle of the table, knowing this is the last opportunity to pull it off.

The first re-test went 97% well, we had 10 dishes to re-present, and that is what brings me to be stood in a car park of Lingfield Racecourse at 4am. A cold late September morning, with a service station coffee in hand. I had left Doncaster at midnight in my hatchback, crammed with cool boxes and ingredients. I wasn't taking any chances. I had prepared the final 10 dishes at Doncaster, driven through the early hours, even calling at the artesian patisserie supplier to pick up an overnight batch of fresh products. We were this close; I wasn't prepared to get stuck on the M25 or have a supplier not turn up or anything else go wrong. I wanted it done, and I wasn't taking any chances.

After all that designing, testing, re-testing, starting over and scouring the UK for the very best products, the final 10 dishes received the sign-off. We were done, after almost a year of work. We had sign-off. We now knew what we were cooking, for during what would turn out to be the maddest summer of my life.

Summer, Greenwich Park, London 2012

I spent the first half of 2012 finding chefs, spending time at suppliers, working through the ordering, purchasing small equipment and working on what was to be termed the "bump-in".

Most things in 2012 went relatively smooth, planning, site walks, working out logistics routes, hiring members of the project team.

As well as been interviewed by what turned out to be MI5, on whether I posed a threat to national security and food terrorism. That wasn't an insignificant afternoon at work either.

We began the off-site food production, where for 4 weeks we had a dozen production chefs, pre-cook 50,000 staff meals, ready to give us a head start on the workforce catering. The scale of the whole Olympic job really started to hit home at this point. 50,000 meals! Even when you say it quick, is a massive number and a very sharp reality check. The production was done in one of Doncaster Racecourse's production kitchens, and blast frozen, sealed and then shipped to both Greenwich Park and Eaton Dorney, following a colour coded labelling system.

At the start of July, the production was finished, meals were shipped and we had begun to bring the two sites to life.

Then the plans changed.

On a wet evening at Eaton, we had been following the news that the contracted security company had struggled to recruit and fulfil enough security posts for all the games. The British Army would now be deployed to replace the majority of the security posts. For a couple of minutes we had a joke, "at least it wasn't the catering that had f*!ked up" went the chatter.

The next morning, I was over at Greenwich. "Tom, how easy is it to increase the numbers of workforce you have planned for?" The whole plan for the workforce catering was devised based on grams per person per meal, multiplied by the expected workers and volunteers on-site each day, give or take 4,000 per sitting, four sittings per day, seventy days straight.

"Why?" I, asked? "Well, you know how the Army is now manning all the security posts?", "yeah". "Well, they have demanded to LOCOG, that the portions of the available food be doubled!" "So we need twice as much food than originally planned!"

Here is when I think I lost it! Almost two years of planning, costing, testing, re-testing, ordering, pre-production and logistics had just gone through the window. What do they say about the plan being the first casualty of war?

I hit the phones! Advanced all of the deliveries up by two days, "can double the order on everything against the workforce orders! and can you have it to me in the morning?"

We were now producing 16,000 portions of food rather than the original eight thousand, across the two venues, that's not 16,000 portions per day, that's 16,000 portions four times a day, for the next 60 plus days. Inwardly, I was freaking out, but I knew I had to stay calm; the chefs at both venues were looking at me for leadership and reassurance. That single scenario took me right to the edge of despair. Throw away all those plans. I had to just go with it.

The opening ceremony, Friday 27th July 2012

As Paul McCartney sang "Hey Jude!" in the Olympic stadium, during the night of the opening ceremony, I was driving around Greenwich Park, dropping off supplies, fixing gas bottles and making sure that service equipment was in the right place. We were pushed, really struggling to get finished. We kept going all night, moving and shifting all type of food. We carried on all night, we just had too much to finish. I had chefs at it all night, some went back to the digs, and others stayed. At about 6:30am I decided that I needed some sleep. I grabbed one of the sleeping bags we had for emergencies, a pack of oven cloths and laid on top of a pallet of A10 tins at the side of my desk in a marquee kitchen in the middle of Blackheath.

I woke after about 40 minutes. It was the opening day, 70,000 ticket holders were arriving at 9am, along with VIPs, athletes and workforce.

I was still on a mission, this time top of my problems was a box of very bespoke morning tea canapés for the VIP's. How to find a box 30cm by 30cm in all of this chaos?

That wasn't my only problem, Environmental Health had just arrived, 6 of them. The power had tripped in the 16th Century former Royal palace. The retail concessions didn't have enough bacon. The workforce needed more fruit. Athletes needed extra chicken; it went on. "It was a complete shit storm".

We managed to convince the 16th Century palace to let us cut a hole in a window and run a three-phase generator cable through to the makeshift service area. We sorted the bacon and chicken issues. It was just the matter of making sure we were one step ahead of the EHO. By step ahead, I mean making sure we sent a runner round the back with a spare mobile hand wash station. Running between those units which were a little "light" on some equipment.

We saved the tea canapés just in the nick of time, to make sure Will, Kate and the rest of the Royals, enjoyed their Cross-Country events.

But it was a crazy day, we got beat, we ran out. 70,000 people can eat and drink a lot. It wasn't just us. The chocolate people, the fizzy drink sponsor, the ice-cream supplier, they all got smashed. They had emptied the lot. We needed re-enforcements and we needed them quick.

I hadn't slept for close to 48 hours. The thing about a lack of sleep is that it makes you delirious, it makes you lose all sense of logic. After a good night's sleep and some food, I got back on the line and I pushed on. We eventually found a rhythm, we at last had found our feet. We just needed to keep on top of the workload now and keep the food flowing.

As days turned into weeks, we approached the transition from Olympic Games to Paralympics.

We had overcome the delays to orders, we had overcome the shortages, overcome the over orders, the staff shortages, the lack of equipment, the 16th Century marble floor. We did it after all of it.

London 2012 was the hardest summer of my life. But in all, it had consumed almost two years. It was an incredible experience, one which has set me up for the rest of my life. London 2012 and the whole project was something which will never be repeated, certainly not in my lifetime.

All of the people who supported that project and there are too many to list, all were simply top draw. Everyone really did pull together, saved each other and some saved me. Some gave more than I could have ever expected. We had support from people everywhere, people who were not even on the sites. Phone calls and unexpected help from people who now have become lifelong friends, suppliers who made their name and built a business off the back of the PR.

Importantly my wife and my brother. They were both there for me in the darkest of moments, when I needed to cry, needed to laugh. I couldn't have got through it all without them both.

For me, I went back to Doncaster and eventually in 2013, left the catering industry to move into teaching and eventually apprenticeships (that is another story itself). But it was a summer I would never forget.

The Humanity of Hospitality
A Waiter's Memoir
Conor Perrott

The Hospitality Industry. For most people outside the industry, this is little more than a phrase, a connotation if you will. To them, it is a collective image associated with the representation of the pubs, bars, and restaurants, or the food service industry as a whole. And even in some cases, defined by a social construct, an undesirable industry to work in.

But what about those inside the industry? What do they think?

Well, for me, and I believe for many others, Hospitality is more than the industry we work in. Particularly for myself, and many of my close friends who I've come to love through our shared time whilst working in Hospitality, it's a place of belonging. Somewhere we can call our own, a place that lays outside the business of our imperfect lives. A place where strangers become friends, and friends become family.

During my days in college, one of my lecturers asked me if I thought that the industry is dictated by the needs of our customers and that the growth in new trends is organic in origin. At the time, I said yes. But now, looking back, I think that is only partly true. I've seen the industry change drastically within the past few years, which is amazing because hospitality will constantly be in a state of flux – forever adapting. The reason for this? Its workers.

Throughout my journey within the industry, I have met some amazing people. People who are so ahead of the game with their imagination, innovation, and passion that their ideas and drive, bring new custom through their doors. It is because of these people that the industry has soared higher and higher, breathing new life into itself.

It is these people that I want to focus on, the gamechangers.

This will be my tenth year working within the industry and honestly, the people I have had the pleasure to work alongside and with, have been a blessing in disguise. Every time I've thought about moving on, giving up on the industry, they were there, becoming inspirations that encouraged me to stay, to keep going. One thing that shocked me about our industry, was the amount of compassion and humanity contained within it and these gamechangers, are the forerunners of it.

Who would have thought? Not me, if I'm totally honest.

Reflecting on my time within the industry, there were some memorable people that made my time within the industry so much more pleasurable with their compassion and humanity.

For years I have witnessed the verbal, and sometimes physical, abuse bombarded at hospitality workers. And yet, despite enduring it to retain their job, these people still rise above it and maintain their kindness. I found it interesting. Well, I still do, as tt is an amazing feat. I know in my time I've been pushed to the breaking point from my customers – like the time I accidently (I use this word rather loosely) told a table of five women during the Christmas period to "get f*#ked" after making demands that their £62.50 cocktail tree was to be taken off the bill because they hadn't read that their chocolate cheesecake dessert was, in fact, a chocolate and coffee cheesecake. But I am human, what is there to expect? We all have bad days now and then. They asked to speak to my manager and to my profound gratitude, my manager at the time, Dom, dealt with the situation. I was working close by, I could her the ladies' cursing me. I could feel my face getting flustered whilst I was with other patrons, I knew deep down this too would end badly for me.

In my mind every worst-case scenario played through my head. I avoided the table immeasurably, sometimes taking the longest way possible to another table nearby, I looked over at Dom and he was smiling with the guests as they were leaving and it wasn't until after they had left, that I started to calm down. I put my head down and cracked on serving, I knew it was coming, I was going to get sacked. I ended up in the kitchen, scaping some plates when Dom came in, his face was unreadable.

"Can I speak to you at the end of shift please?" He requested.

I mumbled a quick "Yeah," and scurried back onto the floor. I spent the rest of my shift dreading what was to come, and despite everything that had happened, I made a fair number of tips and had a cracking time with my colleagues. But the thought of my chat with Dom was there, nagging at the back of my head, like a disease slowly consuming me. I wasn't closing, and neither was Dom. So, we worked the floor until the end of food service and packed menus away for the drinkers that would fill the bar and our tables. I got changed and found Dom sat at one of the high tables near the garden, I joined him, and he had got us a couple of pints.

Dom asked me about what happened, and I explained everything in great depth, adding voices for effect too. I sat waiting for the dreaded news to come when Dom burst out laughing.

I was confused, why was he laughing?

Turns out, I do some really good voice impressions and Dom really liked the extra emphasis on my version of events.

We talked some more, mostly about work in general, but then he asked me about home. At the time, I was going through a rough patch with my family, my Dad had been sent to prison and my Mum had been diagnosed with a rare condition that effects the vocal cords. I didn't get in trouble that day, and it turns out, Dom saw right through me and knew I needed to talk. It was never about the audacity of the ladies' demands that day, it was the build-up of events that were out of my control. And for the first time in a while, I felt that I had a manager that cared for his staff.

As it turns out, he wasn't the only one.

Natalie was one of the Deputy Managers I worked under during my time in the industry. I will never forget the horrendous period I started working with Nat. The team was in a transition of change – a new manager had started, and everything was being reviewed. I started the second week in November (I know, not the best of times to learn a new menu and drinks specs) and I had to train myself like many of us do in the industry. Fast forward to

Christmas and the chaos that followed was unlike anything I had ever seen!

Imagine, you're completely booked on a Saturday and three out of twelve members of staff call in sick for the evening shift. By the time I started, the floor was a mess and the bar was barely functioning – everyone was too busy, but we were getting by. I remember me and Nat taking the two sections adjacent to one another so she could support me if need be. The shift started how it meant to go on: chaotically. We were an hour in, when I realised, I had not seen Nat in her section for some time, so I jumped on taking orders and trying to keep up on my own tables. That's when I saw her… food stained and drenched with sweat. Nat was whizzing around the venue, her arms filled with plates, I suspected mostly dirty. She disappeared into the kitchen briefly before returning with her arms filled to the brim with mains – I realised then, she had been running a three-course meal for a 40-person booking all on her own, whilst clearing the plates from dirty tables as well. We made eye contact briefly, but we were both swept away by our customers.

A couple hours passed of non-stop work, the sort of non-stop work that you get when you go to take a swig from your drink and some crazed customer screams at you because they are waiting to get served – that kind. Nat and I eventually caught up as the 40-person booking started to leave, we decided to briefly go over our tables to each other just in case one of us got caught up (which, by the way, happened almost every weekend since I started). We were at the peak time for service and again, I was running from table to table, order to order, trying to keep on top of the waves of people that were swarming me. I looked at Nat and I could see a twinkle of wildness in her eyes – clearly, she was having just as much fun as I was. That shift, Nat and I must have not spoken to each other for a good three hours, but we worked. I greeted, seated, and wined and dined our guests, occasionally having to jump on the bar and do my own drink orders. Nat ran food, cleared, and cleaned tables, and in between all that, whisked desserts out of the kitchen to her heart's content. The system worked.

When it came to closing, a brief respite allowed us to collect our thoughts and stared out into what I can only describe as a bombsite. There were glasses everywhere! Nat, Jade (the only other person on shift) and I sat in the garden as the venue emptied out. We sat there, discussing things not suitable for work – the kind of things that were, in fact, discussed at work but just out of ear shot of our guests. It had been one of those shifts that left you wondering why you were sticky in so many places that shouldn't be sticky, covered in unidentifiable substances of oranges, reds, browns and with every muscle left throbbing and aching.

Nat tried to give us a pep talk, to heighten the mood to help make light of the workload, before heading up to do the cash up. It didn't help, not really. There was so, so much work to do. Jade and I plodded along, droopy headed and fatigued.

Half an hour passed before Nat came back down from the cash up and cracked open a few beers. We each grabbed one and gave ourselves a cheer. Nat's phone rang and she then headed off to the front gate returning moments later with a surprise pizza, cheesy chips, and garlic bread! It wasn't much, but it meant the world to us. A simple act of kindness. We left work that 'night' around 7am in the morning – it was the busiest shift I've ever worked to date, and it would have been one of the worst without the pizza, beers, and the leadership that Nat provided. Sometimes it takes the simplest thing to make someone's day.

Another gamechanger is a lady called Lou, who is still a good friend of mine, who I meet up with every couple of months to catch up. She was the first manager I had that believed in me, pushed me to go further and nurtured me to grow. It was the first time I was excited to work in this industry.

I suppose it all started when I first joined her team. It was the summertime, and the pub was pretty busy with the beer garden and the 2 for 1 on cocktails. I was working both the bar and floor at this point in my career, flipping to match the busy areas of the business. There was a particular Sunday where we ended up working together. It was dead – it usually was on a Sunday night. I had begun setting up the back area for a corporate lunch the next day, when Lou came over and I could feel her lingering eyes burning in the back of my head.

"Conor, what do you want to do with your life?" Lou asked.

It was a fair question, what did any 21-year-old want to do with their life? I had no idea. When I was younger, I wanted to be a chef, in college I wanted to work in events. But now that I was in the workplace, I didn't really have any goals. I was IN the industry, what more could I ask for? I replied honestly, and Lou pondered for a moment before going back to the bar. As we closed the business, Lou stayed and helped. We talked about our lives, exchanging nuggets of information, and talking about stuff we had seen in the industry. It was entertaining, Lou was from London and the stories she told were comical, something you'd only see or hear about on a tv show. It made great conversation, so much it distracted me from the fact Lou hadn't done the cash up.

When we finished, I went to head upstairs to sign out and get my belongings, when Lou shouted up after me, "Excuse me, where do you think you're going? These tills aren't going to cash themselves up, are they!"

I stopped and looked at Lou, and it felt like someone really saw me for the first time. And yet, for someone who barely knew me, she took a huge gamble. I was at the mercy of yet another stranger's kindness. That night I cashed up with Lou. It was like I was cashing up for the first time; I had cashed up in previous venues but never like this. Every detail, policy, procedure was talked about in great detail – it was refreshing. As my time progressed working for Lou, she pushed and pushed me to learn new skills and take up new responsibilities. Someone who only a couple of months before was almost a stranger, was investing their time in me.

This is the compassion of the industry: gamechangers who put humanity in hospitality. These are the forerunners of a new generation of workers, a generation derived from the kindness, innovation, and passion that is shown to them. Ultimately, they are the ones that will pave the industry to new heights, just as their predecessors did. One of the lessons that stuck with me from Lou that I feel is invaluable, is that when a guest enters your venue, whether they be drinking, lunching, or dining, we have the power to supply them with an escape from reality – however brief that may be.

The ability and understanding that you, as a server, bartender, or even a chef can alter a guest's mood and brighten up their day for even half an hour is something we should aim for. And this can be said for the colleagues too, because this is what makes the industry a home. After all, isn't that the aim of being hospitable?

A trip to the Culinary Olympics
Steve Thorpe

We all recognise how important professional development is to those working in further education, although for most organisations this tends to be more about learning styles and behaviour management, missing out on the importance of maintaining up to date professional skills and credibility.

In 1996, I was still a newish member of a large specific Food and Beverage team in the Hotel School at City College Norwich, having joined them in 1993, when I was made redundant from the Army.

I had received an update from USACAT (the USA culinary Team in Fort Lee) that they would be at the Culinary Olympics in September which was being held in Berlin with some of the team I had worked with in 1989 -

1991. This started a conversation with my new colleagues who liked the idea of trying to organise a trip for professional development.

I was nominated to speak to the head of school and organise the event. So I started putting together a reasoned argument for 10 staff to take the college minibus, by road and ferry to Berlin for a 3 day weekend, which was no mean feat. Well, I succeeded in organising the trip, otherwise this would be shorter piece. It was agreed that this would be a good activity for the F&B team, and it would be completed before enrolment and the start of the academic term, ticking all the right boxes.

The trip was planned, we had an overnight ferry, with cabins from Felixstowe across to Hamburg and then a drive on what was once a corridor for many service families to get to Berlin. The driving was shared between myself and Pete Hallam, an ex-army baker who had served in Berlin and wished to see the changes in the city.

We set off in good time, but only just managed to get checked in to board the ferry, we settled into cabins and then agreed to meet for the buffet supper that was indicated on the ticket, and we enjoyed a smorgasbord of delights and left the restaurant for a small beer. There was then an announcement over the public address system, asking the leader of the Thorpe party – which was our party - to make their way to the pursuers office. We had no students with us, we were all well over 21 so we weren't sure what the issue was.

The pursuer informed me that we had walked out of the restaurant without paying for 9 meals, I had missread the ticket: the meal voucher was only for the driver, while the others had to pay. Not something we had budgeted for, we were to pay before we would be allowed off the boat. I returned to my colleagues with the begging bowl to collect "dinner money". we had a Scotsman with us who would never pay more than a couple of quid for a meal anywhere, even claiming his wife had only allowed a small amount of cash as breakfast was included at the hotel.

We paid the outstanding bill, half of the group then decided they wouldn't have breakfast as they could get something cheaper once we set off on the road, and we also had some snacks and fruit on the van!

The road trip into Berlin was straight forward, commentary on where the police and Russians would sit, and watch service families travel the corridor during the cold war. Our hotel was close to Tempelhof airport which still looked iconic as we drove past. The hotel was fairly new, close to the trams and included breakfast. Just like that, the school trip colleagues had sufficient for lunch and breakfast. Chefs seem to eat poorly (or ratjer, cheaply) as an everyday practice.

We had 2 days in Berlin and spent the whole of the Saturday visiting the Culinary Olympics, in wonder of some of the display pieces of sugar, chocolate, marzipan, dough and salt along with the hot cookery classes. The collection of materials, ideas and samples that we were going to take back and be able to introduce into our classrooms was outstanding, ad I also managed to meet up with a number of British and US Army past colleagues, and even drink more than my share at times.

We managed to do a little sight-seeing too, getting to key points around the city like check-point Charlie, Brandenburg gate and of course the wall or parts of it that remain.

Our journey back was not quite as eventful, 4 of us split the cost for the buffet supper and were able to get some snacks for those who had spent their pocket money elsewhere and couldn't afford to eat, not that they were starving.

On our return to college, we had to make a presentation to share our learning with others across the school, and, of course, I had to explain to the Head of School what had happened with meal booking errors, for weeks being teased about it.

We had managed a 5 day CPD (Continuing Professional Development) activity mostly funded by our employer, which we felt was a great investment, in my 25 years at Norwich I manged to complete CPD activities in Shanghai, Kazakhstan, California, France, Austria,

Germany and Dubai, all to the benefit of students and organisation, with some pain and gain for myself. The camaraderie of hospitality workers across all areas of the sector is what makes the sector such an exciting workplace, you develop friends for life who will support you through the good and bad times.

Nightmare on Chef Street!
Kenny Child

Royal Air Force school of catering

February 1980

Salon culinaire

Inter station team final.

So here we are the inter station team final.

The rules state that the team would be a senior chef i.e. sgt or above, and a junior chef i.e. senior aircraftsman.

I was less than a year after leaving Great Yarmouth college and a couple of months later was in the Royal Air Force

myself and a Sgt, that I won't name, had been practicing for a couple of weeks and had nailed down the menu and all was good.

There were six teams all having the same menu to prepare and serve in 1 hour.

The menu was typical late 70s and was:

Scampi Provençal with pilaf rice

Veal escalope cordon bleu

Fresh fruit salad

No problem at all, the boss had split the work load so my jobs were, Scampi Provençal, Bread crumbs and pane the veal, and making the fresh fruit salad.

The sgt worked on the pilaf rice, butcher the veal and cooking and garnish.

The workload was a bit unbalanced, but we were ready and off we set for the journey from Suffolk to Hereford.

Obviously, the night before the cook off we assembled in the bar chefs from all over the RAF, and beer was flowing.

My instructor from my 3-week stint at Hereford came over for a chat. "Hey, I wasn't expecting to see you, your name's not down for the junior competition?"

Of course, being a little cocky I replied "of course not, I'm in the big boys cook off"

A few more beers and of course lots of bravado from everyone there.

Fast forward to competition day, and beforehand we were given the option of bringing our own ingredients or ordering from the catering stores, and we opted for bringing or own pans etc and serving platters.

I was in the kitchen sorting our equipment, whilst the boss went to get our ingredients. I was aware that I was by far the youngest of the competitors, but that wasn't a problem for me, in my impeccable starched and pressed whites and hat I was ready to go and win.

Alarm bells started when I looked around at some ingredients that other teams had. We'd decided that our fresh fruit salad would be simple but with a fantastic stock syrup, but it was obvious that the others had bought all manner of fruits.

Anyhow, the sgt finally brought our ingredients and looked flustered. "is everything ok chef?"

The reply wasn't quite what I wanted to hear.

The clock started.

The first issue. We had been practicing with prepared scampi, but I now had to shell and take all of the membrane out.

Next issue we had been using prepared breadcrumbs, but now I had a loaf of bread to work with.

Anyway, I cracked on, the judges were walking around taking notes as our working practices were also under scrutiny.

I'd trusted the boss to check our produce (bad move).

I'd seen him starting the pilaf rice and I had worked my butt off to catch up on my timing, the scampi was in the pan, "where's the wine chef?" It's in the plastic tumbler I grabbed the tumbler and raised it to smell and then it hit me, "Chef!!! It's not wine it's olive oil" so there I was asking other competitors if I could nick some wine, but somehow I cracked on.

Service hit me like a steam train.

"For F***s sake chef, you've used that oil instead of wine in the Pilaf!"

The other teams started to snigger, as my exclamation wasn't exactly quiet.

So we had stodgy, greasy pilaf, he'd over cooked the veal, and we had a fresh fruit salad that was looking a bit threadbare.

The judges retired to make their decision and I was left to wash up, not a happy chef to be fair, but at least the other teams weren't as brutal with their mocking as they could have been.

The judging panel came back and the waiting spectators watched on as a group captain head straight for me with a pint of lager and simply said "I think you need this airman".

To my amazement we got a certificate of merit, must have been a sympathy vote, because the other senior chefs came and shook my hand and told me to keep my head up. The sgt was fuming, and someone asked if we'd be in the bar later, but we went straight back, keeping my mouth shut the entire journey back to Wattisham.

Monday I was back on shift in the officers' mess, trying to keep my head down and deflecting questions. At around 9 am, the catering officer appeared in the kitchen with a face like thunder. He just about acknowledged the chefs as we stood to attention with a "good morning sir", and he went straight into the chefs office and slammed the door shut.

About 30 minutes later he emerged red faced he approached me and simply said "I hear you didn't let us down. Thank you, carry on". That was it.

The story apparently had hit Wattisham before we even arrived back, the sgt never mentioned it but he applied for a posting very soon after and left a couple of weeks later.

The joke of: "can you get some stock made please?"

"Yes, what sort?"

"A laughingstock!", soon faded.

Ghostly Goings On
Kenny Child

Sometime between November 1979 and July 1983 at RAF Wattisham.

The junior ranks mess kitchen was manned 24 hours a day, Monday to Friday, but at weekends it would shut at around 7pm, with a duty chef on call should there be any emergency rations needed for things such as diverted flights.

This one particular Saturday night, I had locked up and handed the keys into the guardroom and was in my accommodation when a knock on the door came. It was two Raf police requesting me to get the keys as both of them thought they had seen someone walking through the dining room. I accompanied them and could find no sign of any disturbance. All doors and windows internal and external were locked, but they were both adamant that they had seen somebody in uniform.

Indeed, quite often, the cooks that were on their own in the dead of night reported cold chills from the corridor to the side of the kitchen, and me included. The keys were returned, but the police were both sure they had seen someone.

Around two weeks later, we were on a station exercise and I was on the night shift. On exercise all ranks ate in the junior ranks mess or had hot box meals sent out to various parts of the station. Us, the cooks, would all be busy in the kitchens, however the stewards would-be put-on guard duty including around the mess.

The mess was two storeys, but the first floor was not used and had store rooms which gave us an idea. We had been telling the new steward about the ghostly encounter of two weeks before, and, as we waited until about 2 am, 2 or 3 of us went upstairs and proceeded to move a chefs jacket with a broom handle through the sleeves across the window, where we could see the guard.

We also rattled a few old pans about to catch his attention. He looked terrified, but dare not leave his post. We were wetting ourselves, until one of us (not me I might add) decided to up the ante, throwing an apple at him. Unfortunately, it hit him bang on the head with some force, and his yell alerted other guards. All hell let loose, as it seemed we were under attack.

Bless him, to this day I suspect he still doesn't know it was us. We knew that if we spoke up, we would be in serious trouble.

But the sight of him with a bump on his head explaining to the guard commander that it was a ghost that had done that to him, was a sight to behold.

We believe that the guard commander knew what was amiss, but considering our boss was in on the prank, nothing more about the incident was said.

Baptism by Fire
John Sullivan

Breakfast Time in an armed forces kitchen
 of old
Lights go on as the breakfast chef opens the kitchen door
Sturdy boots cause echoes as he walks across the red tiled floor
Trays of sausages go onto the Collins oven roller plates
He slides them in with a large hook and closes the gates
Beaten scrambled egg are placed inside the steamer
Porridge also with salt and milk to make it a creamier

Hot salamanders stand ready for the food to be grilled
Bacon Tomatoes and black pudding, trays already filled

Now other chefs have arrived to give a helping hand
Opening the beans and tomatoes that are canned
One brat pan is ready, as slices of bread are placed in
Another is for frying the fresh eggs, all to be cracked within

Poaching pans are on standby, just in case there is some need

Now the dining doors are open, there are hundreds that we need to feed

Orders are shouted to the chefs, who are providing the support

Making sure that those on the server duty, will never go short.

There are also a choice of cereals and breakfast wheats

Others have fresh fruit or toast as their morning treats

Milk is in plenty, so is the urns of boiling hot tea

There are lots of different choices for them all to see

Now the breakfast is over, everything is cleared

Into the Pot Wash all dirty items are steered

Ready for the next excellent dinnertime meal

The chefs are already preparing with professional zeal

.

Them's the Rules
Tony Oxley

From having 25+ years in the industry I decided to make a career change and go into teaching to pass on my knowledge and skills to others who wanted to be chefs.

I knew nothing about teaching, and upon my full day interview, they gave me the job, however, they could not understand why on earth I was taking a £24000 pay cut, why I would go into teaching. I must've been mad.

I had come to the point in my life where I was not seeing enough of my wife and kids, every day off I was spending the day falling asleep. I needed some quality time with my loved ones.

Plus, I liked the idea of a nine to five job and weekends off.

I can clearly remember my first weeks of teaching, teaching theory from an OHP projector, no whiteboard and a flip chart and paper was a challenge, and everyone kept everything they had created to themselves. To find a flip chart marker pen was a challenge.

I was given a class called "Fast forwards", in my naivety.

Thinking this was a group of learners who were being fast-tracked due to their ability. How wrong was I?

This was a group of year 10's from local schools in the area, who may have shown an interest in cooking. In reality, these kids were the naughty ones, who I think had been encouraged to attend college to give the teachers a day's rest, or so they could have a day off school.

I inherited this class from another lecturer, who I am sure was having a nervous breakdown at the time. I was to teach them a level 1 NVQ food preparation and cooking course. It was a group of 8 boys and two girls, who bullied the boys.

I soon learnt I had to have eyes and ears everywhere to have some form of control over these kids.

Even giving knives and hot pans to them made me feel very nervous. I remember the day one student left a plastic bin lid on top of the salamander, which caught fire. I thought I was surely going to be fired.

They knew I was the new boy and would take advantage of this.

These sessions were held on a Friday, and the amount of Alcohol and paracetamol needed afterwards developed two-fold. I thought I drank a lot with my chefs from the industry at the weekends?

It became apparent early on, there needed to be some classroom management, I would've definitely been in trouble if anyone observed my lessons or we had a call from the dreaded OFSTED.

I needed to put some ground rules in place, but I did not want to become a dictator to the group and become the nasty teacher, so I decided to give them some ownership to collate a set of rules and forfeits which would incur if the rules were not adhered to.

Also, there would even be forfeits for me too if I broke these rules.

The students were fully engaged with this initiative, however, I did draw the line at their penalty of having to wear a pair of Y-front pants on top of his chef trousers if I was late to the lesson.

Every student had to agree and sign this contract which was laminated and displayed in the kitchen. I loved the day when 7 students had to sing to a level 1 group of hair and beauty students a song of my choice as they hadn't met the deadline for their homework: I chose the apt song of "Man I feel like a woman" by Shania Twain. There were a few protests, however, I just waved the laminated sheet in front of them to confirm they had signed it. I learnt early on with this particular group that if I showed any sign of weakness, they would exploit it.

Give an inch, they would take a mile.

The swear box used to fill up quite quickly early on, until they realised they were losing money and believed they did not need to be the Gordon Ramsey's of the college. The fines for missing uniform soon built up, and those funds went towards resources for teaching. (Came in very handy as the government kept cutting FE funding).

I especially liked the rule for disrespecting the tutor: the perpetrator became the chef's servant for the day.

People may find some of these rules cruel, however, they gave some excellent life skills like being organised, learning to respect authority, learning to work colleagues and as a team, to do things like clean down the kitchen.

After about two months we had productive lessons, where classroom management reflected what was expected within the industry. Getting them ready for work was essential, and also understanding that college was not like school.

One lesson I learnt personally, was when I was late, my forfeit was to buy the whole class two fun-size bags of chocolate bars, and to give it to them at the end of the day and not during the lunch break,. I made that mistake one day and it was like they all had three cans of Red Bull each. I was back on the paracetamol and alcohol that night. They even had the cheek to tell me to remember there are no mistakes only lessons. Boy, that was a big lesson learnt that day!

One thing I can say is that I never succumbed to the forfeit of missing my deadline for marking, so you will never see me singing on "The Voice" or "Britain's got Talent"!

Throughout my teaching career, I have always found that setting boundaries with all my learners, showing respect and having fun, creates a perfect place for learning and being creative.

Exactly what I would like to see in any chef coming into the industry.

Head Hunted
Kenny Child

In early 1984, I was a young corporal in the Royal Air Force. My first posting after promotion was at JSMRU RAF Chessington in Surrey, and it was the joint services medical rehabilitation unit, for injured servicemen and women, a fantastic posting for me, and 30 minutes from the centre of London. I was young to be in the position I was holding, which meant that some of the older cooks were suspicious of me. They quickly realised that I was a decent chef, so all was good.

One afternoon I was called into the catering office, and I was told that I had to go to Raf Innsworth, in my best dress uniform to attend an interview. Innsworth was the admin centre for all postings.

What's the job then?

I was told that I was up for the position of personal chef to the Air Officer commanding Strike Command Raf High Wycombe, which would be a 3 year posting and I'd be basically promoted to Sgt at the end of it.

I was pleased that I'd been recommended, however this wasn't a job that I wanted: cooking for the family of this officer would entail doing high profile dinner parties.

My issue was that I would effectively be working for the Officer's wife and I would be at her disposal whenever she wished. I would not really be part of a brigade, so it wasn't for me.

The day before I drove down to Innsworth and went to my room for the night, I made my way to the mess for dinner and chatted to the chefs on duty and arranged to meet for a beer later.

In the bar, chatting with the lads, as usual they had their fingers on the pulse, "so you're looking forward to the new posting then, nice cushy little number?"

I replied that obviously I only went for the interview, which was met with a grin followed by a "good luck mate".

The next morning I was sat outside the office in my sparkling uniform waiting for what I hoped would only be a 5 minute chat. I was also aware that there didn't seem to be any other candidates.

In being called in, I was quite surprised to see a low-ranking officer waiting with tea and biscuits. I sat down, and realised it wasn't an interview at all: I was read out a list of responsibilities and basically told that my posting would be in around two weeks' time and if I had any questions.

I stated in no uncertain terms that I would not accept the posting.

The officer looked flustered and obviously not impressed, and I was simply told to wait outside.

Around 30 minutes passed, and an admin clerk came and asked me to go with him to another office. A more senior officer was waiting, and a similar conversation ensued. This time I told him that I believed that I would be better served elsewhere.

This scenario happened once more, and I was in front of another more senior officer who basically told me that if I didn't take the prestigious job that indeed I could say farewell to any promotions, and it was likely that I would be posted to an area at the other end of the country. They knew that I had bought a house in Norfolk, and thought that the threat of moving me to, for example, Scotland would sway me.

Now, when serving in the forces you go wherever you're told and whenever it's needed, however I also realised that they would not want someone that didn't want the position, and the officer that I would be working for probably didn't even know my name and if it didn't work well, the guys at Innsworth would take the flak.

So back to Chessington I went in the knowledge that a promotion wouldn't happen anytime soon.

However, within 12 months I was posted to Norfolk and stayed there for the rest of my time.

Did I make the right decision?

Perhaps not, however I needed to be in control of my own life, so knowing that, I think I probably did.

Chocolate leaves, has that young man lost it?
Kenny Child

We used to do lots of formal dinner functions, rather grand affairs, with silver cutlery and silver service. One particular dinner I was in charge of desserts, and I cannot remember for what dish it was, probably a mousse of some sort, but I decided that I wanted chocolate leaves: for 80 covers I needed 240 perfect leaves.

Now bearing in mind this was early 1980s, I told the head chef that a tree on the drive had perfect leaves we could use to mould the chocolate leaves on, so off I went in my chef's whites with a ladder and a bag. I'm up this ladder, selecting leaves, when I noticed a few strange looks from officers going past. An officer cycled past and stopped, came over to me and asked what I was doing. He was the station medical officer, so I explained that I needed

240 perfect leaves for the dinner the next evening. He asked if I was ok up the ladder, and was on his way.

He went straight to the kitchen and asked if they were playing a prank or had I been drinking. They assured him it was my idea , but he gave me a strange look as he went past.

Fast forward to the dinner. On these occasions, it was traditional that the senior chef went out and share a glass of port with the senior officer present. On this occasion, when the sgt came back in, he told me that the station commander wanted to see me. So out I go, escorted by the head waiter, as I was greeted by applause and a pint.

It seems that the word was, one of the cooks was not mentally stable and was up a tree on a ladder in chefs whites.

By all accounts, the station medical officer got a bit of stick when he saw that the sweet was garnished by perfect chocolate leaves.

Salmon on the parade square
Kenny Child

I cannot substantiate this tale, however I heard it during my time in the RAF more than once so it is very likely true.

There was a VIP lunch being held from the mess for some reason, and the centrepiece was a whole salmon poached and decorated with aspic, as was trendy back then.

The salmon was set on a large mirror and had to be transported across the parade square, so two chefs were carrying it and at some stage disaster struck the mirror cracked and the salmon ended up on the concrete, inedible.

The Raf caterers are nothing if not resourceful in time of adversity, so the cooks grabbed someone to clean the mess up, and hurried back to the kitchen with the head and tail, and numerous tins of salmon. They placed it all into the mixing bowl, to be bound with softened butter and shaped to resemble a whole salmon covered with finely sliced cucumber.

Everyone was happy with that meal, apart from the Station Warrant Officer, who by all accounts was going mad at what had happened on his hallowed Parade square.

Be Careful, Chef, That Knife is Sharp
Kenny Child

So way back in 1994 I was the chef manager of a 13 bedroom hotel in the picturesque village of Walberswick on the Suffolk coast. One Friday evening my fish delivery was late, so I was trying to catch up with filleting and prepping at the same time as the early part of service. Back then I was a fan of Global knives, as I loved the balance etc. Any how I was filleting some fish and somebody called "check on Kenny". I was distracted for a milli second, and turned just in time to feel the blade on my thumb. It was immediately obvious that it wasn't a minor cut, however there wasn't much blood, so I duly wrapped an oven cloth around it and proudly managed to serve mains to another 2 or 3 tables before feeling a bit wheezy and sat down at the edge of the kitchen. The bar manager who was decent at first aid, had a look and said "Jeez, you've cut the tendon!"

One of the locals in the bar took me across to Southwold Hospital where they had a minor injury unit. The doctor on duty was a GP and stated that it would be ok if he glued the cut so I let him carry on without him realising that I was holding it in place with my fingers underneath. The devil in me just couldn't help myself when he said " that should do the trick". I laughed and said " I don't think so, doc" and duly moved my fingers and of course with the tendon cut my thumb flopped open.

I was packed off to the James Paget hospital in Gorleston. The lad from the bar drove me up, bless him, so I arrived in A&E still in chef's whites. Although it was now about 8:30, it was thankfully quiet, so I was seen quickly, a surgeon came along and said he would be able to repair the damage with local anaesthetic.

A couple of minutes later I was waiting, but could hear the conversations going on as I heard a radio conversation "road traffic collision multiple casualties one suspected spinal injuries police attending with suspected drink driver". The surgeon came back in and asked if I'd heard all that and said that they would have to patch me up and asked me to return Saturday morning.

So the next day with thumb bandaged and secured with a metal tube I returned. A different surgeon came to examine me and told me that unfortunately the tendon had regressed and I would be admitted for an operation under full anaesthetic.

So Saturday afternoon I went for the op and I was to spend the night in hospital with intravenous pain relief and antibiotics. Now, I'm not the best patient, but some young lad was getting on my nerves with his constant moaning and being rude to the staff, but the real crunch came just before lunch when I spotted a police officer talking to a nurse at the end of the bay. Of course I, being nosey, was listening intently. It was when the officer went to the lad's bed and said "right son, the good news is you've no damage, now I'm arresting you for driving whilst under the……"

I was straight of the bed dragging my intravenous drip along, raging at him for being the reason I couldn't have been dealt with earlier. The officer politely advised me to calm down and return to my bed. Whilst this lad was getting dressed to be carted off to the station, the officer approached and warned me that my aggressive manner wasn't required, but he understood how I felt and smiled and shook my hand. A few minutes later the doctor came and looked at the mark on my thumb, saying I could leave but wondered what I had cut it with as he said the cut I'd made was cleaner that the two with his scalpel.

I had to return a week later to have the stitches removed, but by the Tuesday I was in the kitchen running the pass one handed.

The Colourful Linguistic Challenges, be Careful What You Wish For, Tendewberrymud

Steve West

Pranks, pranks and more pranks are talked about before going into a kitchen. Some more obvious than others, like the classic long wait when sent from a chef who is your leader and a person you are supposed to look up to.

Other ones include, "don't be scared when the lobster being sent to its demise screams".

Clip clopping through the kitchen on the first day as an apprentice, is a very scary thing to be. Gleaming whites blinding the other chefs in the kitchen, my clogs a claxon ringing out, NEWBIE COMING THROUGH. Knives in the canvas wallet, every knife, placed in its correct slot, can't have the chopping knife going into the office knife slot, RRIPP, oops too late.

What the hell is this small curly one for? It was the 70/80s, and it was used a great deal.

I had a rota of sections from Garde Manger, which means keeper of the food, but is in fact the larder section, to butchery, sauce, the king of sections, and entremetier, the vegetable section.

What wasn't apparent at the time was these were not stand-alone sections, these were highly linked to each other, dependant on each other to come together at any given focal point, the high value of customer expectations.

One the sections I was to be tasked with was the Coffee House.

This was a 5-star luxury hotel and the clients were premium, from princes, to those who wanted and craved the service of pampered style of a millionaire lifestyle.

The Coffee House in Park Lane was an extremely busy kitchen with tourists, room service and, as an indicator, across the road from the Buckingham Palace,

Piccadilly street and a stone throw from the Ritz, Hard Rock Café. The area was a constant vibrant, colourful, exciting place to be.

Possibly the world's richest square mile in the world, overlooked by the Hilton and around the corner, the Gavroche.

Even with all this going on, my job as an apprentice was to do as I was told, keep my head down and just get on with it.

Working the Coffee House was our lovely workers from overseas. Generally from the Philippines and Thailand, both in the kitchen and front of house.

There is music in languages from other countries. It is a great way to learn different languages, cultures and it enriches one's life in the industry.

However, on entering a kitchen, over the noise of the extractor fans, the checks being shouted out, the clanging of the pans, one must also develop an ear for the orders which are for you.

This is a skill one must learn.

But, and this is big but, working with peole who speak different languages, come challenges from all sides.

In the auditory tidal wave of commotion, we have a Filipino, a Thai chef and a Londoner trying to communicate to each other on top of the throng of yet again Thai and Filipino waitresses shouting through the hotplate.

Now, it would be rude of me to say anything other than I do not speak either language and I had the utmost respect for them.

We had a great team: Sam from Thailand and Larry from the Philippines took me under their wings and helped me a great deal.

There was an onslaught of checks and orders from the restaurant. The ear of comprehension was starting to be developed.

One tiny little detail had to be explored.

Sam and Larry could not pronounce 'V's, and neither could the waitresses, so that letter was replaced by 'B'. My name now turned into Steeb. Brilliant. As if that was not enough, I was on the grill and fryers and on my side was the Veal or Veau Viennoise.

The barrage of checks started to escalate.

Steeb, 2 Beel, Bienwoias. De bow.

The waitresses lined up and shouting my name was like being in Trafalgar Square with car horns blaring.

I could only hear: "Steeb, Steeb!"

During a quiet period, Sam, Larry and myself were taking and it occurred to me to ask why they could not say the letter 'V'.

They just could not and the following conversation happened.

They tried:

"Sam, STEEVVVBBB."

"Steeeeve."

"Stib."

"No. Steevvve."

"Steeeebbbb."

"Steven."

"Steeebben."

We just cracked up and had to concede. The thing was, they knew it now and almost begged for the whole night to filled with BEAL BEINWHAS, de BOW.

Waitresses would shout out, and of course understood the crack now.

They changed pitches for a laugh and my name was sung out, "Steeb, Stib", echoing like a hundred and one car horns were going off in sharp bursts.

On this particular night we were exceptionally busy, the service started to throb, orders were coming in as quickly as food was being expedited.

I hated the grill, and I hated my name as well. Burgers were fine and my ear was firmly in sync with that.

Right in the middle of service I was right in the middle of Beal and Burgers, I shouted over to Larry, as I wanted to exploit learning a different language, I thought this would be a great opportunity.

He came up to me and looked over to the waitresses, five standing at the hotplate, eyes, and eyes ever watchful as the food was being cooked.

"Mate, how do you say, 'go away' in Filipino?"

I followed his eyes as they were nattering and he whispered "Suput Ako".

Now, the fact he whispered it should have been a giveaway.

My chest puffed up, I had leant a different language, hadn't I ?

Here was my chance, I poked my head through the hotplate, and they were all listening as I looked around and proudly stated I had learned some new words.

Arms flaying and beckoning them away, I shouted in fun, "SUPUT AKO".

For a split second a massive foreboding overcame me. There was a great pause and deadly silence except for a comedic clank of a pan moment, and in my thoughts, I sincerely hoped Larry hadn't stitched me up.

Within a brief moment, they all burst out laughing and a massive phew entered my head, and a brow wipe. It must have been my accent, so I thought.

After a couple of weeks, this "Suput Ako" seemed to amuse them. There was a lingering thought though, as deadpan Larry looked over to me and winked with a knowing smile.

There was a break in the service and one of the older waitresses came over in a hilarious state.

I asked what was up, and she asked me I I knew what I had been saying over the last couple of weeks, and of course I thought it was "go away".

I looked over at Larry, who was in stiches, looking over her shoulder in the distance.

Well, here was the rub. All this time I have been shouting out to all the waitresses, "I am not circumcised!"

There was a momentary feeling of horror, but from there on to the future, my name was not Steeb anymore, it was now,

"HEY SUPUT 2 BEAL, BEINWHAS DE BOW".

The lesson to take away from that is be careful what you wish for!

My dad was a chef himself, trained at Westminster Culinary school in 1954. He worked in many place, including becoming Executive Chef of Fortnum's and Masons for a while.

After a stint in the UK he decided to move to the United States as a junior sous chef at the LA Marriott, then proceeding to be the General Manager at Souverain Winery in Napa Valley. He then moved to the San Francisco Yacht club where he entertained Expats who settled there. Of course, this was right up his street and his humour was that of The Goons, Spike Milligan and those comedy gold moments which in his words, the Yanks didn't get.

He emailed back and forth stories that usually were hilarious however, a lot could not be repeated.

As true stories go, there is an email which highlights a time when language was a problem for dad, when he was working in a hotel in Istanbul.

It tells of a time when he ordered room service and recorded his experience.

It goes something like this.

Subj: TENDEWBERRYMUD.

Date:04/16/2000

From: Wineman

To: Loads of people in his list.

In the content, dad mentioned this had to be read aloud.

(Footnote, from myself, this nearly killed my spellcheck for ever. SW)

Room Service (RS)

Barrie West   (BW)

RS:"Momy. Ruin sorbees"

BW: "Sorry I thought I dialled room service".

RS:"Rye- ruin sorbees, morny. Djwewish to odor sunteen?".

BW:"Uh – yes I'd like some bacon and eggs".

RS: "Ow July den"?

BW:"What!"

RS: "Ow July den? Pry, boy, pooch".

BW: "Oh the eggs! How do I like them? Sorry, scrambled please".

RS: "Ow July dee baychem – crease?"

BW"Crisp will be fine".

RS: "Hokay, an son tos?"

BW: "What !?"

RS: "San tos. July son tos?"

BW:"I don't think so".

RS: "No. Judo one toes?"

BW" I really do feel bad about this, but don't know what, "judo one toes" means".

RS: "Toes! Toes!, why djew Don Juan toes? Ow bow singlish mopping we bother?"

BW: "English muffin!! I've got it. You were saying toast! Fine.

RS: "We bother?"

BW: "No just put bother, on the side".

RS: "WAD?"

BW: "I mean butter, put the butter on the side".

RS: "Copy?"

BW: "Sorry?"

RS: "Copy, tea, mill?"

BW: "Yes please, coffee and that's all".

RS: "One Minnie. Ass ruin torino fee, strangle ache, crease baychem, tossy singlish mopping we bother honey sigh and copy- rye?".

BW: Whatever you say".

RS "Tendewberrymud".

BW: "you're welcome".

Mealtime
John Sullivan

Steaming kitchens with stock pots simmering steady

Blanching chips in brat pans for frying, almost ready

Pots and trays are filled with various savoury's and sauces

From the talented chefs of our proud Armed Forces

Roast meats are in the ovens basting

The Corporal's are giving all a final tasting

Making sure all is up to standard and ready go

Shouting out orders so all the team will know

The cooked vegetables are ready and comes off the steam

Everything is like clockwork , it works like a dream

The carving boards are down, to place on the roast meats

The pastry chef has prepared and cooked all of the sweets

The WO's and Sergeants are like the leaders of a band
Conducting their chefs and giving an experienced hand
Garnishes are been prepared to finish these delightful dishes
Presentation is important as is their orders and wishes

Hot plates are now waiting as chefs whites are been changed
They know who is on the servery it has all been arranged
Soups of the day with bowls ready, is first for the taking

As outside the doors a crowd mumblings of anticipation is making

Disposable Art, that is the thing that these chefs have created

Many diners are noisy and some have patiently waited

Now let in, many do enter, the hungry women and men

Soon it's all gone, the chefs have to start all over again

Don't Take the Pee
Kenny Child

In July 1983 I was bussed from RAF Wattisham down to Brize Norton to embark on a four-month detachment to RAF Stanley. The flight to ascension was relatively uneventful, although standing on the tarmac surrounded by armed guards at Dakar airport was a bit disconcerting. While we were being refuelled, once at ascension we were flown by sea to the SS Uganda, the ship was used as a hospital ship during the Falkland Island conflict, and now used to ferry us, the troops.

Luxurious it wasn't, the food was not great, and the civilian caterers were more used to being in the Mediterranean as Uganda was a school education cruise ship. The only other person I knew was an Air traffic controller that was on the minibus with me from Wattisham, so we ended up in the same dormitory with around 20 other airmen. I nabbed the bottom bunk given the fact that we would encounter very rough seas, and I thought that being on the bottom was safer with less distance to fall. Around night 3 of a 10-day journey, disaster struck. We had been allowed alcohol for the first few days until we entered what was considered an operational theatre i.e. Within reach for attack from Argentina, and, as you can imagine, the bunks were for young school children, they were not comfortable with very thin mattresses.

    I was half asleep wrapped in my sleeping bag when I felt something warm and wet dripping on to my head - instantly I guessed what had happened and was straight out of bed. I made the decision to keep quiet as I didn't want to wake our entire room, so

        I went straight to the shower and cleaned myself up. I then spent the rest of the night on the floor using kit bags as a mattress. Next morning, as the guys were waking up, someone asked why I was on the floor, by this time the culprit on the top bunk had stirred as I replied, "you'll never believe this, but some dirty bastard pissed all over me".

There was much laughter until matey from the top bunk exclaimed "bloody hell whoever it was has pissed on me my sleeping bag is soaked!"

He kept well away from me for the rest of the journey, and in fact I only saw him once on the island which was strange as I was working about 50 metres from air traffic.

Our Story
How Not To Kill Your Partner Working With Your Other Half
Reena and Henal Chotai

It all started on a blind date on a cold winter's day in January 2012, and a common love for food and making people happy. Fast forward 9 years and here we are. Married for 8 years, have an 8 year old cat, 5 year old café and a 4 year old daughter!

Working in any business/job with your partner/husband/wife is stressful enough, but throw that mix into a hospitality business and it's either a recipe for disaster or something magical. Thankfull, we fall into the latter category. We're both in the kitchen, 7 days a week, 10-12 hours a day and surprisingly, we haven't killed each other... yet!

The past year has taught us a LOT, and more so how to appreciate one another. When we are in the zone and during a busy service, it's almost like a dance that we do in the kitchen. No words are spoken and the music in the background is the beat to which we work. Plates are ready to receive food, sauces are drizzled, toast is buttered, chips are drained, omelettes are flipped, burgers are assembled... it's almost the perfect dance.

We have our bad moments. If we didn't, we wouldn't be human! Yes, there's shouting, the occasional scream and a heck of a lot of swearing, but what happens in the kitchen stays in the kitchen. We don't bring work home or dwell on it. If we did, we wouldn't be able to survive, both in our marriage or our business.

Communication is key in the kitchen, not just shouting 'BEHIND' when walking with something dangerous, but knowing who is doing what and when. The orders on the rail, the timer starts, the target to get food ready is 10-30 mins depending, on how busy it is, or what dishes are ordered. We are in sync, and have a system for which we each give a time then adjust our cooking to match up. Any dish sent out contains our heart, passion and love as the secret ingredient. That is one thing not ordered or made, it runs through your body and empowers you.

We work in a stressful environment, with pressure on timings, heat, where you are on your feet all day, working unsocial hours, and with dangerous equipment everywhere. Top that off with working with your other half and you'd think it would be a recipe for disaster, especially when working with knives, pots, pans and hot liquids. Nothing has happened, not yet. The focus is on the ticket. We know what we each have to do, but we have one eye what the other is doing, we have each other's back. We work together to plate, as a team.

It's not always easy and perfect, we each make mistakes occasionally. The swearing and name calling starts when we have to redo a dish. If anyone from the outside heard the way we spoke, they would say "That marriage is not lasting till the end of the day". Far from it, once the dish is sent, the words said in the previous minutes are wiped from the memory and we move on. This is one of the superpowers we have to have to stay Mr & Mrs.

Our happiness is not dependent on finding out the contents of the till at the end of the day, it's dependent on the empty plates coming back or the great comments from the customers: this is the best feeling in the world, and it's the gold medal we strive for.

The kitchen and the café are an extension of our home. We love to entertain and to make people happy with food cooked with love. To have had to suddenly stop everything on that fateful day in mid-March 2020, after running at full hilt, was like a punch to our stomachs. It hit us a few days after we closed, when the realisation of how damaging this was going to be crept upon us, but not just for us, but for the whole of the country. It was almost as if the world stopped and paused, and someone out there was waiting to push the "restart" button. But it gave us the opportunity to take some time to spend with our then 3 year old daughter, something we never did as we were always working.

We took time to appreciate the small things, like appreciating our home, the surroundings we live in, the beauty of nature, our marriage and our lives. Material possessions became almost unnecessary, when we had our health, we had a roof over our heads, we had food and we had each other. This is what taught us to be grateful for what we have and to help those who may not be as fortunate as us.

Thinking outside the box, we set up an online delivery service to provide basic daily essentials to those that couldn't get to the supermarkets or get an online shopping done. This enabled us to keep in touch with our beloved customers, whilst keeping ourselves busy. Through this, we were able to donate food and snack items to our amazing frontline NHS staff at our local hospital (a special place for us as we were all born there). We were approached by customers, family and friends who also wished to help but know how to, so a GoFundMe page was set up and the donations came flooding in.

The fundraising page has been phenomenal: the generosity and kindness shown by our local community has left us feeling really humbled and in awe of the kindness of strangers. We speak as one voice and our community knows us as one, rather than as 2 people. The Marcus Rashford campaign for free school meals, which we participated in, propelled our drive to lengths we didn't know we could reach.

As broken and as fragmented the hospitality industry was forced to become, the virtual world opened up a new way of connecting people and the hospitality family. Connections and friendships were made in a short space of time. Advice was given from the country's leading chefs, hoteliers and we were given a chance to voice our opinions and fears on the same platforms as these great and inspiring people. Between us, as a couple, we had our set tasks to carry out and whilst it was exhausting, it was incredibly overwhelming and humbling, knowing that we had the support whenever we needed it form the wider hospitality family.

It may seem like we have the perfect life: working together, running a business and a home. But it's incredibly stressful. We have moments where we fight, argue, disagree, shout, swear, slam doors on each other… but we have a common goal: to help those who need our help.

We had a slight scare when Mr ended up in hospital just before Christmas, but thankfully he is fine and on the road to recovery, but it did make us stop and think that our health has to come first now (it was always on the back burner), above everything.

So, this is our story – the abridged version! Hospitality is our life: we love what we do and making people happy with our food and service. We serve food… with a purpose.

And to those out there that maybe think they cannot work with their partners/husbands/wives etc…. it can be done. It's bloody hard work, but it's incredibly fun and rewarding at the same time.

End Ex End Ex
Kenny Child

It's a regular occurrence that operational Royal Air Force stations would have a Taceval (tactical evaluation) or exercise. These would normally start with a blast of the air raid siren and everyone would be straight up and at their work place as quickly as possible. These would last for anywhere between 12 hours and 5 days.

All of the catering would be done from one kitchen where we would be operating both in house for the old equivalent of click and collect. Hot meals would be sent to the front line in hot box containers. At Raf Wattisham we would feed around 2500 people over a 24 hour period.

Us caterers would work 12 hour shifts and it always wound us up that when an exercise finished most would head for the bar but us caterers had to carry on with no let up.

One taceval went down as one of the biggest examples of a prank gone wrong and as usual, the chefs were at the centre of it. Communication was normally over a tannoy system which would broadcast to most units and of course that included the dining room.

So, during one very busy taceval, we were particularly short due to overseas commitments, so by day 3 we were shattered and we were having lunch by ourselves in the mess. We had an in-house tannoy system, and one of the corporals thought he would raise morale.

"Stand by for broadcast, standby for broadcast, at 1500 hours local we have successfully repelled both air and ground forces and base continues to be held secure, furthermore I would like to thank the catering squadron for their magnificent efforts over the past hours in particular the egg banjos this morning were superb end ex end ex". We were wetting ourselves, as his impression of the station commander was spot on, however what we hadn't realised was that several guard commanders were still at the far end in the coffee area, out of view. We saw them only when they stood up to leave. Disaster was about to happen as they believed that the message was correct and began to stand the guards down across the station, less than ten minutes later we heard the message: "Stand by stand by all guards are to return to post immediately I repeat all guards are to return to post immediately".

Less than five minutes later, a senior officer emerged and it was fair to say he was raging and demanded to know who the culprit of the prank was. Needless to say, we shut up shop, and nobody said a word. He threatened to have us all put on a charge, but of course this fell on deaf ears and he stormed off. The only fallout was that the taceval was extended by another 48 hours. To be fair, the Catering Officer knew exactly who it was but we heard that privately he found it funny.

***

Over the past few weeks while writing these tales, the memories of some really great times in some of the many kitchens that I have performed - and I say performed because hospitality is like being on the stage every customer is a critic – come to mind.

Back in 2019, I'd fallen out of love with catering, with the long hours over many years, the changing times when some companies treated us as numbers, and how our usefulness was limited to producing great results on the trading reports. So in September of that year, when the doctors advised that I should leave the kitchen behind, I was ready to accept it with no issues. However, I was put in touch with Hospitality Action, and during the next few months I was continually touched by the kindness of many in our industry, and during the writing some of my tales and subsequently becoming an ambassador for this fantastic charity, I have made some fantastic new friends that otherwise I would not have been in touch with.

I will not name people because those that know me will understand that my stroke brain may make me omit someone that I wouldn't wish to.

But in a nutshell, we have we have the best people ever in this trade of ours and I thank you for making me love it again.

I would like to feel that I have one great service left in me, maybe we can get a brigade together for a swansong?

But for now, and me, it's End ex End ex.

Getting My Family Barred from a Restaurant... The Walk of Shame
Steve West

Customers tend to think they can get away with trying to get money back, which they think they should be entitled to whilst the chefs, waiters and bar staff are only trying to make a living.

There is a blight on hospitality people: customers who believe the person in front of them at that time is at their sole service. There are times of course when things do go wrong, things happen and within reason people complain, ask for assistance, and it's our job to assure the customer's experience is memorable and enjoyable.

However, there are occasions when this does not happen. As a chef and assessor who has been in the industry for over 40 years, it boils my blood when things do go wrong, and I am given lazy excuses from the kitchen or management. Never say you are a chef, or work in hospitality because then you have lost. What would you prove from this, what is your goal? To make you feel better. We are all in it together as caterers and this has to be respected.

Being a family man, I love taking my family out in family-run establishments, or brands which offer simple meals and nothing challenging.

Never stating about working in hospitality, complaining is difficult, but on occasions it really has to be done. The restaurant sign *please wait to be seated* welcomes us, the family hungry for, gammon, sausages, mixed grills, a 5 star restaurant wouldn't suit them and frankly would be money wasted, but at the end of the day family time is very important, where ever it is spent. This does bring about problems, because it ends up with us deciding not to go out because some establishments simply do not care enough.

There were several occasions when the food did not meet the required level standard, never mind one of being edible, as the trolls toenails served were disguised as onion rings, sausages undercooked and when going to the manager to say something, I noticed the chef overlooking from the open plan kitchen from the grill, spying me, deciding to come over to take my said sausage. Nope, didn't want excuses, just a sausage which wasn't as pink as the flesh from the day of its birth. It dawned on me, when spying on the chef on his phone across the same kitchen, that he was probably researching 'how to cook a banger without killing another human being', but I doubted it.

The subject of chips came up as the family dug into their fare of the evening, and it was very apparent a breeze block could have been cooked better and thoroughly on this particular occasion. I'm not one for complaining, but why is such a simple form of cooking so difficult? It's a chip, a spud, an inanimate object which has no soul, no life, but if it did, it might have done a better job if it had grown legs, peeled itself and plonked itself in the fryer all by itself.

The obligatory question was asked, 'How was everything sir?' whilst of course your gob is full of food, and you are unable to answer at that time, maybe that's in their policies? In a paragraph 10, subsection B, to ask the customer a question while they are eating. If said customer nods and is still breathing, we will take the feedback as Brilliant, 5 stars.

So, the family looked on, not happy, but trying to bite through golden brown scrabble pieces, waitress saw the eyes, her face didn't show much expression, she knew there was something up, and started to walk away until our 'excuse me. Er the chips aren't cooked'. Knowing full well these were not checked and simply cooked too quickly, it wasn't a big deal, however, money is still being paid and I just wanted my family time to just go nicely.

'I will get the manager' she replied.

A youth, aka the manager, came over and in true Basil Fawlty style started into my eyes, as he seemed to bend at an angle of 90 degrees. Quite impressed by his athleticism and flexibility it still didn't take away the fact the chips were not cooked properly. Now, as a first world problem, it is not that big a deal, however, when a transaction is being made in return of services, that is a problem if the said service has not been delivered correctly.

Understanding the chips needed to be cooked and a wait was imminent, the next part was legendary as the youth, aka manager, started to reel off all of his best excuse as to why the food wasn't cooked properly. It wasn't like we wanted it brought to us on a carnival horse or clowns transporting in a broken car while tooting their big noses, on gold tinted tray with glinting diamonds. We just wanted our chips!

'Well sir, unfortunately this time of year we have had an enormous amount of rain which has affected the crop from the farmers and…'.

This onslaught on my ears had to stop, and I explained that I did not care about the farmers name, his wife's name or incredibly enough the size of his shoes, the name of his offspring because all that was required on the evening family event was properly cooked CHIPS!

What bemuses me is why make excuses, other customers did not care either, and here is the point: all anyone wants is cooked food on a plate and a question time debate is really unrequired for the entertainment of the evening.

Our chips arrived, cooked, proteins gone, but never mind eh?

There was another problem but dare not even go there, the peas were rock hard and obviously microwaved within an inch of their ex-green pod life.

All good, it is only peas, after all.

Another round of drinks were to be collected from the bar and who should be there, but the previous son of Basil. Drinks trayed, ready to go as my thirsty hands clasped the black tray and then an image of a pea entered my head. I just couldn't let it go but not wanting to spoil the evening with the family and out of ear shot, 'by the way', his ears cooked toward me, 'the peas were rock hard too'. There are moments in time when, no matter how hard one tries, the build-up of such a simple thing will simply not go away.

'Frankly' annoyed and wanting this off my chest, 'it would have been quicker if I had sent my boy to Iceland, come back and put a pan of water on the stove and' – at this moment the manager grabbed the tray in front of quite a few people and spouted, 'YOU'RE BARRED'.

'Don't worry about your bill you can leave sir'. I didn't want to leave.

The walk of shame took me to my family where I had to mention to them, we have been barred. The word 'we' took the burden from my shoulders. In front of customers, we walked the plank to the outside of the restaurant and heads bowed, upset, only because we missed a mountain of dessert and ice creams. As it happened, we went to Iceland and loaded up on the delicious sugary overload to soothe our pains of disappointing experience when it didn't have to be.

The Cook, Soldier Moth, EHP and the Groin Grabber
Steve West

Customers who frequent our restaurants and establishments on the whole are the best. People who frequent and revisit places where comfort is essential for the enjoyment of an evening, a day off from the daily slog of life.

Customer demographics are part of our game. Cosy couples, excited parties for birthdays, weddings, Christmas, - where the Turkeys get off lightly, events, stadiums, burger vans, take-aways, fish and chip shops, hotels, bed and breakfasts, seaside towns hoping for sunny weather and rain to not stop to play out: this what we do, who we feed and the nature of the customer varies from place to place. Other types of customers include conference-goers, where the delegates stay over for a few days, on the company cash. These are generally training events, CPD that can have up to around 20 people in a group. The norm is, a delegate would get an allowance, a day rate for food, no alcohol of course, the price of the bedrooms and breakfast.

Tallied up, this can surmount quite a price, if the event runs for several weeks and totals up to be a pretty penny to a company.

If a company runs a course for 12 weeks, this can mount up to thousands of pounds.

A particular hotel I took over as head chef ran such events. Lucrative for the hotel's gross profit, these were no brainers and once the menus were put into place, it was just a matter of ordering and controlling the situation. Fundamental really, one would think.

When starting I was introduced to the Food and Beverage Manager, George. We chatted, talked about where we worked, who we knew, as we tried to find a common ground, and we did. We clicked. An initial meeting was set up to go through the procedures, policies and the real meat on the bones, the gossip, who was doing what to who, and an idea about customer expectations in this particular hotel.

Being warned, George told me about a particular conference company, who were always trying to get a refund and usually found a way to get their money back for a reason or another. The heads up was invaluable, and being vigilant and a control freak, I had to create my own system to put into place. First call was to the local authority for a cuppa, a chat and to build a relationship. This was always the first thing on the list, to build a two-way trust between us. Over a time, he would come in and discuss things. Lovely chap always helpful and there for me anytime if I had any problems. My restaurant manager, Dave, was really good with customers and knew his wines, which the customers loved, as typically they do not like thinking too much when out dining. Orders would come in, and if there were any problems, he would sort things out. There was one teeny little fly in a particular ointment, the smallest of things which anywhere else would go unnoticed, except for the fact hospitality people have an extreme heightened observation radar. That is, do not do anything in front of a chef who will see it, only to later bring it out for everyone's amusement. Many examples can be given but too much as I am sure you have your own examples.

Dave, as good as he was, had a tick, a habit which went unnoticed to himself, and it was an unfathomable, inexplicable pinching of the crotch of his trousers .These incidents were not a one off, a now and again thing as his thumb and forefinger in slow motion to us would clamp like a crab claw, travel southwards and slowly manoeuvre the bottom of his clothed zip and pull away from his manhood.

George and I discovered that the more excited, he got, for example talking about funny stories and chatting about, he would clamp more and more. We would call him over in times of boredom to have a competition to see who would win the prize of the coveted "knob scratch of the day award".

The bulge snatch would repeat itself, and stress could induce a spasm of pink flesh claws grabbing at his flies.

Checks and paranoia were my rock in keeping tabs. A rolling 7-day checks of customers order system made sure these would not get one over on the kitchen. Sure enough, a customer felt ill and used the Food Poisoning card. Let's see, yesterday, Coq Au Vin. Strange? This was called off the specials board before they came in. They tried sever times to get their money back.

It was the last day of the conference for a few weeks, and we managed to dodge their bullets of trying it on.

Lunch was sent, delegates sat eating and the event managed to go ahead without, so far, any incidents. 'Fag time George?' 'Sure, let's go'.

We gathered the nicotine filled cardboard packets from my chef's office and with one foot ready to embark on our nasty inhalation habit, the most savage blood curdling scream came from the restaurant door into the kitchen. George and I thought someone had been stabbed, beaten to a pulp or even worse, someone could've seen Dave smashing his hand against his groin area.

The screaming got closer as we came back into the kitchen, where we met an ashen-faced waitress holding a plate of food in her hand.

The plate looked untouched, and trying to get any sense out of the bawling waitress was futile as she was trying to gasp for air. George took the plate and placed it onto a table in the stillroom. In unison, we looked at the plate, turned to each other, back to the plate and together mouthed 'What the fuck is that'?

I thought George was a bit harsh on the Lasagne, but on the same plate as the Lasagne was, new potatoes, mixed leaves and various salad items garnished with what looked larvae.

George and I just gaped, and to be honest, we almost shat ourselves. A calm head was needed, as I sent for Dave who was on his break. Dave came in as this needed to be dealt really carefully, as it was the demon conference from hell, and words had to be rehearsed to make sure we were not giving any money back.

After soothing Dave down from his Bavarian crotch dance panic, he had to be assured this was not from this kitchen.

My new friend, the local authority was called in. Meanwhile the evidence had to be dealt with. The conference organiser banging on about having a complete refund for the whole 12 weeks had to be told they would not get an answer until it was fully investigated.

The plate was wrapped tightly, labelled and frozen, which didn't need it as the local authority sped down like a Sweeny Police car, screeching at the gates and made a speedy entrance.

His eyes lit up, and you could tell he couldn't believe his eyes.

The conference leader seeking answers became a pain in the neck, calling George daily, the whole issue was escalating, as the General Manager wanted this mess going away. It took 3 weeks to get an answer from the local authority. At last, but he wanted to come down and speak to me face to face. He first addressed the larvae, which is what it was, and was being kept in the Science laboratory in Cambridge, because it was the Larvae of a soldier moth never seen in this country before. He went on to explain, in front of myself and George, that the larvae would have been cooked and we wouldn't have been able to recognize it, if it was in the lasagne, and we would've seen it when cleaning lettuce, if it was on the salad. However the larvae was on the side of the plate. Dave rocked up, eyes wide, crotch grab and listened. As far as they were concerned there was no reason to refund the customer. He then asked us who the customer was.

It was a horticultural training course.

Thank god it wasn't zoological training event! Couldn't bear to see a penguin tilted on the side of a dish. Needless to say, being paranoid about that particular customer trying to cheat us, was worthwhile, and we got it sorted in the end.

Here is a lesson for complaining customers: think about those behind the scenes trying their best. It doesn't always work out for you in the end.

Kitchen, Service, and Shouting Orders for Dummies
Steve West

Food emporiums can, in many ways, be paralleled to theatres, even though the script of the kitchen menu is in place, the rehearsals are not the same each day as the chef lives in the world of momentary "If anything can go wrong... it will, with moments to spare, just as James Bond with seconds to go, saves the world", but also ovens not lighting, or working properly, kitchen porters and chefs not turning in, food not turning up on time. Today's excuse? Because of an accident... again, in a faraway fairy tale land where delivery drivers live which is usually in a lay by somewhere, texting, sleeping, or trying to think of a really good excuse why a particular, special item of food ordered by the chef, for an important dinner party, is not in the back of his van ready to be prepared finely with passion for his guests.

The result of this slight oversight could induce an anger so rare, the chef would just leave the back door with a copper pan implanted in the driver's skull. The Mise-en-place or preparation, composing of menus setting the stage in the culinary theatre with precision placed irons, earthenware, glass and linen are being assembled ready for the curtain to rise around 7pm for the evening performance.

Kitchens are hot, loud, noisy caverns especially during a busy service. The rising tempo of clattering pans and the intrusive hum of the extractor fan starts to get louder dragging the fumes, steam and heat away from the poor boys and girls in the engine room of the kitchen, battling against time in as calm a manner as is humanly possible under such conditions. If everything goes to plan, tradition should be upheld with the starter followed by the main course and hopefully finished with a dessert.

The foodie Captain at the helm has planned his gastronomic course, momentum gathering for this daily voyage about to set off and jolts of heart thumping, electric adrenaline shooting through the stomach, as a reminder that maybe he should have taken the plumbers course when he had the chance, because in about two hours there will be nowhere to run or hide from the onslaught about to ensue.

The check has been dispatched into the kitchen and the journey begins. Even if one of the waiting staff that brings the check into the kitchen was Mother Theresa, she would still be hated for those words... "CHECK ON CHEF!" Arms flaying, and quicker than a beat of a butterfly, you work to ensure sustenance for the belly while enticing the taste buds as the countdown of time dwindles away.

The verbal method for the information on the check to be delivered to the brotherhood in the kitchen is, in time and honoured tradition, shouting and bellowing, articulated, interjected and strung together with a couple of colourful expletives, of course, and this is then acknowledged by each individual section relevant to an item or items of food roared into the kitchen, er, requested.

The edible picture is then constructed onto the earthenware frame, precision and deftness is needed as it has to pass immediate inspection by the chef who devised this fine puzzle of a one hundred pieces. This does depend of course on which restaurant you enter, from Chef Van Gogh, Constable or Dali on a bad day, hung-over depressed or both.

The Chef Van Lunatic starts to open his noisy orifice, the one where food goes in and the check is about to be bellowed. Now this next little bit is for those fine young people who may find themselves in a kitchen for the first time. Before was mentioned about articulation in the art form of shouting out checks in the control of the service and the listening to what is being said, well forget it, it will never happen until you get used to it further down the line.

So, for those first timers, and poor bastards who enter the kitchen for the very first time.

Presenting....

"The Art of Shouting out orders in the Kitchen and the

Art of Listening... For DUMMIES"

Start with a close friend to practice with. The thing is to try and start slowly because there is no peeking at the checks once they have been shouted out by the keeper of the paper line of orders. You see the thing is, trying to have a sneaky look at the checks, because your mind has gone blank in the middle of service, and you have 101 things to remember in a split second, you are in the shit and the snake in your stomach, medically termed as a lower intestine has grown spikes, can magically cause the vein in the temple of the holy lunatic one, to expand which in turn can induce an anger of magnitude of a scale of... well actually there is no scale. You could, however, on not training one's ear to listen, be kicked out on your arse to a MC'arvesters kitchen. You see in MC'arvesters everybody is allowed to look at those checks and they are proudly placed on the same hooks where the microwave instructions are kept, right next to the colour coded scissors.

There is no getting away from it, one will have to listen. There have been real tried methods to get around this, though over time, and here just for you who are thinking the light bulb above your head is the holy grail of getting on in the hellfire caves, the methods that be thought of but never tried.

Method 1:

Pen and paper secreted in a place away from the Head Chef.

This methodology though simple creates its own problems. The method should be that, when the order is shouted out by the chef, one would take the pen, write down on the paper, what item/s would this lovely person want, and how many. Simple eh?

As the service starts, you would be only able to do this once. This "once" will the first table, and if you are lucky, you only have one item to do.

YOU WOULD NOT BE ABLE TO WRITE QUICK ENOUGH TO LISTEN, BECAUSE AS YOU ARE WRITING, THE KEEPER OF THE PAPER LINE HAS ALREADY SHOUTED OUT TWO MORE TABLES! Ergo, you will be in the shit!

Examples for you follow:

Good news, there are syllables, bad news; the words sung out are not decipherable. The items shouted out will be akin to shouting through heavily woven camping socks whilst wearing a gimp ball strapped over the pink lipped noisy cavern and beware though, this does not take away the actual noise and decibel super sonic boom heading towards one and your ear hole.

The tones and pitch are the secret.

OnGNOrDa!!   This means, On Order, or

SAHMARCHE Ça Marche

In the above example, the emphasis is on the syllables.

Carry on with simple techniques and examples.

SAHMARCHE! No

CHHONGG!	Check on!
WUNG!	One!
NoSTAAHA!	No Starter!
PA...AY...CAANAR!	for this example, the

answer is after the exercise.

For this exercise, get a good friend, one who will not take the piss, blindfold yourself. For that added kitchen effect, burn a bit of toast, tune the radio to a hissy hum, turn the base up and not that you would you look a complete prat, but this is about as close as you could get on to your first day in the kitchen...oh yes, and if possible try to do this exercise on the hottest day of the year or failing that, find a sauna.

First exercise:

The scene:

Imagine yourself standing opposite a hungry Rottweiler just about to rip a piece of your body which keeps the human race going and absorb the adrenaline Niagara Falls rushing around your veins. Get your, maybe ex-friend at this stage, to shout out.

CHHONGG! WUNG NASTAAH! WUNG PAY...AY...CAANAR!

The response should be...WEESHEFFF! This will comfort and sooth the head shouter into his safe zone, and promotion is in sight, well until the next check comes on.

PAY...AY...CAANAR is Pate de Canard. Not to be confused with CONNAR! If you hear this, don't come back into work the next day, especially in French kitchens.

The lesson can be a hard one to learn.

The first check is bellowed and at this point a slight jolted panic sets in because the first order may be yours. Stomach churning apprehension sets in. Head down as the adrenaline starts to kick in and the mind goes into overdrive, busy in your head with thoughts colliding as rehearsed timings and numerous questions are thrown at you from the ether. Have I enough portions, are the garnishes ready, looking around at your section and mise-en-place, jargonised meaning if it isn't ready, hell hath no fury like the chef on the pass waiting 30 seconds than he is uncomfortable with and the discomfort is 10 fold as the feeling in the kitchen mirrors your hell and your peers are so happy, (understated) it isn't them. Don't worry lads and lasses in the culinary engine room of life, your day will come.

During the busy working day in the kitchen for a chef, the preparation of food has to ready for that short period of service time. Food, garnishes, pots, pans, spoons, plates, sauces, fridges cleaned, section tidy, toilet paper ready, check list completed. All this has to be and ready to go immediately when requested. The curtain rises for the performance, the audience is starting to arrive... will you be ready? Time is of the utmost importance and every kitchen in the world has a clock, (except for my local takeaway), the swinging regulator unstopping, and ticks away as a guide to making certain tasks are met, jobs are finished or finishing, food is cooked or cooking precision like, but most importantly, it is make sure one has a enough time for just one more coffee and cigarette before the offensive begins.

This seemingly innocent timepiece is in fact a horrible, living breathing thing with a sick sense of humour. Clocks like to play games with chefs; their favourite one being "let's play catch up". This really is a game one does not want to play. This is where the clock perversely looks down and when your eyes averted unaware, gives a wry smile and adds an hour on just for a laugh watching over as you look up thinking all is ok, and sees your eyes bulge, mouth drying, ear bleeding, screaming internal panic come over you. It then becomes your turn when arms are driven by every sinew in you to drive yourself like a formula one piston engine to get yourself out of the shit which may be around sooner than you think.

The game begins and becomes a very tight and tense race to the winning line. Tempus Fugit, his trophy, your neck on the line, stripping you of all dignity, driving your self-esteem into the trodden ground in one fell swoop and all done in the space of the longest thirty long seconds of your life. And just to lighten this up, in a 16-hour day, it is a lot of 30 seconds.

The head chef has the order in his hand trusting, assuming, (not the safe comfort zone you want to knock out of the head chef's equilibrium), your section has actually finished preparing everything for the night. If all you need is a few minutes to finish off a couple of jobs, this will seem like a lifetime if that first order is yours. This is not the game of catch up you want to play.

As a digression, if any of you who have worked in outside catering, the feeling and consequence of the sickening feeling of the clock is increased by tenfold. The scenario being, all food, starters, mains, deserts, garnishes etc are fed into a refrigerated van, the function of a wedding, two hours away in another county in a marquee on a sodden field, and if by a stroke of luck the van may not get stuck. But more importantly the most memorable day of two people's lives is in your hands.

The permutations of the things that could go wrong are very real. Assuming the van gets to the location, setting up kitchens is needed to be done, food taken out of the van… carefully, there are no shops nearby that would be able to cope with 200 portions of smoked salmon mousse, exotic mixed leaf salad, balsamic dressing and pesto. This also includes the fact one may get lost and hit 5 counties because you had missed a turning you were supposed to get off 10 minutes before.

That is also assuming the marquis has enough electric and water points and the supplier of the equipment hasn't wrapped all said equipment in enough cling film to cover the Isle of Wight 3 times over,

with a collection of trestle tables, splinter free one would hope, as you have to build the damn kitchen first on a hessian, 45-degree angled floor in a middle of a field on a freezing cold day in the middle of winter. Other little factors include, when the party of 200 people are sat, everything is ready to go and assuming all the planets have aligned to a point of synchronicity, sauces are on, condensation is building up inside the busy circus tent only to look up. The ad hoc instruction of the day, "fuck sake Steve, make sure you count those snails crawling along the ceiling" as everyone darts about covering all the food ensuring the customers don't get an extra portion of uninvited proteins in their meal. The marquis has no clock.

A panicked feeling sickening cold feeling comes over you as you glance at the clock, because a fleeting look is all the luxurious time you have, as the timepiece from hell is giving you an evil smile and the finger.

This feeling can only be paralleled to freefalling from a plane, realising that you have just forgotten your parachute and it is way too late to go back and fetch it, as five counties of this green and pleasant is hurtling towards you. The only way out of this would be to knit yourself one on the way down, because you know the evil wrath from all the demons living in the underworld is far more affable than the abuse you will receive from the chef if the food isn't in front of him at the time specified.

Here arises the problem. In one split second there is a roller coaster of emotions, especially in passionate kitchens where the chef eats, breathes and lives for food to the point of obsession. Passionate to the extreme where excessiveness is an extreme, teetering and balancing on the edge of sanity, nerves raw and exposed to the chilly wind of art and culinary performance to achieve that ultimate starred Oscar.

In my early days, in the kitchen, I thought demi-glaze was the grading of eyes that peered at me throughout the working day. The fully glazed, opened, bloodshot, eyes, cushioned with a puffy lovely blue underline, describe a being where one's brain had been dissected from all senses and feeling. The telling sign being, gritty sand grinding adhering itself to the inside of ones eyelids; when actually it was rock salt put there by the head chef while you were taking a quick nap in the changing room between shifts, saying to himself, "that'll keep the little bastards eyes open for a few more hours".

The peak of the roller coaster is in sight, chef points his lovely endearing face in your direction. Nerves shot to pieces, you put into gear, and pistons arms are going like a piston engine, trying very hard, not to show how much of the shit you are in. But now, your new boxer shorts bought from Marks and Spencer may soil at any moment, as your sphincter has hit overdrive and about to explode, but you stand your ground, stomach churning as the hundred geese flying south for the summer in the pit of your gut, chef shouts...

    2 PATE DE CANARD

    2 NoSTAAHA!   Pay attention!

2 FISH DE JOUR

1 POT AU FEU

1 BOUILLABAISSE MAIN COURSE

4 POMME ET LEGUMES

Boxer shorts have averted the turtle head tango and are unsoiled as by chance you have what he wants. You flip your finger back at the clock as you construct, build and send, until the next order.

There is no other feeling in this world. You have been saved, the minutes are there, and the gods have decided in your favour.

The service is now in momentum, in unison the brigade reply... YES CHEF!

The music of the kitchens begins, fridge doors open and close with hushed whistles. Pans land with dull clanks on stove tops and gas rings' as numerous, swift hands dip into their little pots of edible paints from stainless, shiny palettes, squeezy bottles of drizzles magnificent precision, pea shoots, placed on with tweezers, plates run over with a cloth as the clean, pristine canvasses are created as the pressure starts to gather.

To feel a kitchen in full flight in all its magnificent electric glory cannot be imagined, only lived.

To the untrained, virgin eyes, it's just a kitchen with all its cacophony of clanging clatter, but confusion is music to the culinary conductor as the checks are read as the baton is raised as the kitchen orchestra moves as one building music from the sheets with precision, timing and accuracy of the utmost passion which to an outsider may seem like bedlam layered on top of confusion - and at times can be - but if you live it, it can be the absolute sweet spot for knowing you are alive.

Hierarchy: A Lesson for the New Apprentice
Steve West

On entering the kitchen for the first time as a new apprentice in gleaming whites, you can be sure you will stick out like a sore thumb. The reason: your crisp pristine whites will shine like a beacon through the kitchen. Do not worry, this will not last long.

Throughout history, chef's whites used to be different depending on the grading of chefs: the uniforms used to differ in coloured buttons on jackets, on the height of the hats, and on how loud one shouted. But in modern times there are other signs to show the hierarchy in the kitchen.

The first thing to know is that the word *Chef* actually means chief in French.

The noticeable grading of chefs in large hotel goes something like this. The following order encompasses the title of the job, the attire, and the libation of the related title. Unlike now, whites and uniforms had to be bought, washed and worn by the chefs themselves. The grading and state of the whites generally was an indicator of the amount of work and tasks during the working day. This also determined, according to the pay scale of the jobs, how many jackets one could buy. The amount ranging from the Executive chef having two changes a day, to the apprentice who managed to save enough money to have one jacket for the month, which was rewashed daily... through the night, and after an 18 hour day of work... This was called in the trade, a great learning curve, and character building. If lucky, the jacket was dry before starting a day's work.

Executive Chef (big Chief):

Pristine whites (so white they looked like they had a halo around them), sunglasses were needed to dull the glare.

Bottles of champagne

Head chef (small big chief):

Gleaming whites

Glasses of champagne, other hand is kept free for crossing oneself to the big chief.

Senior Sous Chef (under small big chief):

Whites slightly creased due to getting one of the other chefs out of shit.

Small glass of champagne - free hand please see above.

Junior Sous Chef (under, under small big chief):

Smattering of food on apron because finished job senior sous chef couldn't be bothered finishing because champagne came out.

Transition from bottled beer weaned onto small glass of champagne. Chef de partie (Section chef) big chief on little kitchen area:

Smattering of food on apron and jacket.

Bottled beer.

Demi chef de partie (under section chief big chief little kitchen):

Whites are soiled slightly.

Glass of beer and packet of crisps.

Commis Chef (section gofer and whipping boy to all the above):

Bits of white showing through soiled and greasy grey whites.

Sniff of bottles and running index finger inside the glass for the last dregs.

The Apprentice chef (whipping boy to everybody, including the kitchen cat):

A walking compost heap with ivy growing down his back, jacket held together with blue tack, safety pins and cling film. Chaffed, raw thighs from trousers not being repaired on the inside. Knees of trousers worn away while backing out from the Executive Chefs office and again crossing himself.

Manual tasks of fetching beers and glasses for champagne and the last one to leave the kitchen, because he couldn't get out of the large Biffa bin quick enough, while searching for... anything to drink.

This wasn't any real world. The first weeks of apprenticeship were a real eye opener, and any other orifices of the anatomy which had involuntary reflex action at when being shouted at.

Ordinary people finished work at 5 o'clock and went out drinking or home to families and loved ones. Friday night the world stopped until Monday morning at 9 pm when, complete with Monday morning blues and some hangovers, everyone regaling and swapping stories from the previous weekend. Well, that was what my so-called careers officer told me. What he didn't tell me when I took on my apprenticeship that I would age 40 years in 3 weeks.

In fact I was loved so much during my time as an apprentice at the Intercontinental Hotel London, working hard at being amazingly clumsy, and being conscientious in trying at every opportunity to dodge my Executive Big Chief the world distinguished and highly respected Mr Peter Kromberg as much as possible.

As a footnote, an apprentice has no human rights lawyer to aid him once entered into the vacuum world of the kitchen; I believe it is described as character building. On one particular day while playing "dodge the boss", my job that morning was to collect the stores from the storeman to bring back to the Larder section.

It was a huge responsibility because the stores were so far away from the main kitchen that going back and forward to collect items, food and supplies for service, was time wasted and that depended if the storeman was in a good enough mood to give the food and items straight away.

Just to explain, in massive hotels and kitchens, across the world, the buildings are so big and complex; they are actually like little villages connected by a labyrinth of rabbit warren corridors. As a guest in one of these hotels, the facade being well looked after with a daily 24-hour vigil of industrious people ensuring the stay is welcoming, warm and comforting. From night engineers, to housekeepers and chambermaids quietly changing bed linen and origami the toilet paper into the shape of an aeroplane cone, they work, with professionalism and quiet pride. What isn't noticed is the underbelly of this wonderful, complex machine catering to the guests needs, wants, to ease their stay.

Large kitchens have sections. And sections that are part of this machine are the sauce section, vegetable section, roast section and coffee house, which was a little self-running kitchen by itself. This did not include another separate staff kitchen feeding the army of in-house staff and last but definitely not least the banqueting kitchen. Banqueting and catering for large numbers is an art in itself. The logistics, ensuring quality, quantity of hot beautiful food served to 400 people is extremely hard work. Communication and trusting a team is of the utmost importance, both back and front of house are paramount, something that is definitely missing in some establishments these days.

The most incredible service ever witnessed was one in this very kitchen. The Executive Chef Peter Kromberg having the guts, skill and dexterity to serve 400 guests a course of soufflés, each one risen to perfection, as if each one was prepared individually and lovingly tendered.

The jewel in the crown of this hotel was the Soufflé section, which has achieved a Michelin star and quite rightly so, but maybe not so much for the food but, the sheer front in having the balls to create such a concept in a busy fine dining restaurant. The dreaded, dedicated soufflé section has however been the downfall of many a burly hardened chef, and some have come away in terror as the pressure built and another tearful chef had hit the dirt.

Bases had to be perfect, ovens calibrated, tools at the ready, usually a towel and a bucket in the corner of this boxing ring. Orders on, egg whites whisked to perfection to order, one of the ovens open, two were needed to keep up with checks that came on. Wooden paddles turning the soufflés ensuring they rose straight and true. Waiters standing by, with silver platters, ever ready to transport these golds, nurtured edible fluffy pillows to the waiting recipient. Absolutely, being an understatement in this case, no mistakes.

The soufflés could be anything from a starter, main course or part of a main course, or dessert, so timing was imperative and extremely crucial. Needless to say, I never got a look in on the section, especially when I managed to smash a £600 Robot Coupe blender, the day before the chef was supposed to come back from his holidays. I did learn however, that there were so many ways and different languages 'stupid dumb twat' could be said.

So when collecting the stores for my particular section (the larder section, well away from the main kitchen and brigade I might add, where classically, all the buffets for functions, preparation of starters, cold food for restaurant and room service are made), massive trolleys are used to transport the food.

The responsibility of the Executive Chef with aid of his Sous Chefs was immense; ensuring not only the quality of the food was consistent and earned accolades, but also striving to make sure the customers in the Coffee House could have the best Chicken Caesar Salad money could buy.

The stores where all the food was delivered had to be kept and looked after with the same organisational precision streamlined operation as the kitchen, which was overseen by the storeman. The Chef de Partie of the larder would write out his extremely long list of wants from the storeroom, and that particular day happened to be a busy one, preparing for a function which would take place in 3 days' time. Chef de Partie needed someone responsible. He looked around. I was the only one in the vicinity. He looked around again. Head bowed at his list he looked up shaking his head with slight swelling of tears in his eyes.

Here lay the problem: the person who had the list to collect the food had to get it countersigned by the Executive Chef and no one else, as nothing could leave the stores without it. Another problem arose because there are things that you are not told and one of those things, which is a big secret, and one has to learn for oneself is the "work it out for yourself son" lesson.

Even a drummer boy leading from the front had a better chance of survival drumming his army into battle rather than incurring the wrath of not only the Executive Chef but also the Chef de Partie, if you did not return with what he needed, which by the law of all things that shouldn't happen to a decent human being, was normally what the Executive Chef did not want the Chef de Partie to have.

The Chef de Partie of the Larder section would hand over the list to me and just as I was on the starting block quickly passed a comment, "I need this urgently"! This was not good because the chef didn't do... urgent. I would now have to find the Executive Chef for the signature. Like a runner with an important message of life and death I would leg it, well as fast as one could in a place where running was not allowed. "Find the bloody Chef"? He could be in a million places.

I found the chef and handed him the list. He would ask me why I needed the item and why didn't I get it earlier. The problem with being an apprentice is that nobody tells you anything. So there was I with a list of a couple of items which, not only did I not know what they were, but also not knownig why I didn't collect them in the first place from the original store collection, which was given to me by someone who should have known he needed it in the first place.

Here lay the quandary. Arms behind my back, not really paying attention and looking around, I tried not to feel the sense of what was about to come. The chef would not sign the sheet of food items. A hushed whisper of Beelzebub deeply giggled in my head. I had to go back and let the Chef De Partie know. I could just leave the hotel and never come back and sometimes to this day I wondered why I didn't. What I didn't realise, and it was conveniently missed off the 'get me out the shit list', and also it had been neglected to inform me, was that as soon as this list was passed to me it became my responsibility, my problem, my possible demise and soul shredding experience. The Chef De Partie, a large German chef, brilliant, loyal and so committed to his job, was a true Trojan in the battlements of the kitchen, a real warrior. His normal hours would be from 5 in the morning until 12 at night 6 days a week. His face was a skull, sprayed on pinkish skin, dark infills under his eyes that mirrored his striving passion to be the best chef de partie there ever was. He had no lips, as they were dried up and scaly from the dehydration of the daily onslaught.

"Working stuff out myself" mode had to kick in. The thing I did notice about these citadels of the culinary war zone was a need to survive under any circumstance. Kerching! Light bulb had been lit. Sex... Yes sex. Before being an apprentice, I had worked in housekeeping as a linen porter for a little while waiting for a placement in the kitchen to become available and one of the things learnt was where the guests used to discard their... ehem... leisure magazines bought from Soho, obviously to view to alleviate the daily grind, so to speak, and drudge of staying at a five star luxury hotel in the richest square mile of London. And the more affluent the guest, the better the quality the illustrations were.

The magazines were always stashed behind the cabinets in the area of the lift before the guest would depart, or sometimes left in the rubbish trolley of the chambermaid. One had to be careful though, simply because as these were a once only use item and one had to assume the guest had "cracked one off" before leaving for the day so trepidation was usually the word of the day. I would then expedite the contraband, unsoiled of course, and swap it for a beautiful club sandwich during my time in housekeeping. Like hawkers in Oxford Street, ever watching, and unknowing to the Executive Chef, the trade had to be discreet. I opened my coat like a bandit in the night selling stolen watches." Nnnniice"! The deal had been done and a time was agreed to collect the food.

 I had made the call to a friend of mine who was still working in housekeeping to... fetch some premium gear as it could be a matter of life and death. The circle of life had begun. My food passed over for the smuggled goods, which as then passed over to the storeman, who then passed over to me the item of food needed to get me out of the shit with the CDP and everybody was happy. Well for that day anyway, tomorrow was another day. I escaped that day... But

 ...the next day it was to begin again. Having made to believe by the Executive Chef that I was the missing link, this next day I proved him right. Trundling along the lengthy kitchen playing the daily battle of "Dodge the Chef", while walking fast and not paying attention where I was going, I had in my care, the stores sheet, signed this time, and a massive 8 foot long 2 foot wide trolley to put my wares on to. Darting my eyes to the side through the hotplate, watching the sea of chef whites on the other side I knew I was clear, until I noticed that the trolley had come to an abrupt and crashing stop, as I nearly lost control and the trolley nearly tipped over, closely shedding the loose metal shelves and with all the strength I had, retrieve to it back into the upright position, in unison with a muffled Swiss German accent trying to explode expletives to me. The trolley had 4 shelves, and a metal back, but because of my stupidity quota, which I had more than my fair share, I didn't turn it so I could see through the slats.

I turned to look around the trolley to notice the Executive Chef nearly becoming a victim of the glass shell. One could only describe the feeling of when an airplane descends very quickly and the panic of dropping from the skies to impending danger came over me, as the Chef, maroon faced... paused and it was apparent in his eyes, he did not know whether to pick up a knife and hysterically run it through my spinal cord, or batter me senseless with his boot. I do believe however it was the utter disbelief or even shock, that he vocalised his thoughts in front of everyone, and all of a sudden, I understood the name of the part of a women's anatomy... in five languages.

After a while, I put my notice in and was ready to move on, I approached the Chef and he could see me coming through his glass shell that overlooked the kitchen.

As I handed him the piece of paper, one I am sure he would sign off, he looked up at me and I could see he was filling up with emotion, choked up, I could also see the heavy weight lifting from his shoulders and a lovely pink colour return to his cheeks, as he said "about time too"! Wasn't that lovely? I thought as I walked away from his office. I heard the door close behind me, and I am sure I could hear muffled, hysterical laughter from behind the glass shield.

Peter Kromberg did care about his workforce, even coming into the kitchen on New Year's Eve to shake all our hands and wish everybody happy new year. Peter Kromberg spoke 4 languages and he gave each chef the respect of greeting each chef in their own mother tongue. He even spoke to the likes of me, and yes, he said something to me. It wasn't apparent until later because he spoke in a different language to me, while he shook my hand... very tightly. After speaking to Sam, the Thai chef, Franz the German and Xavier the Frenchman, it was broken down into, "Do the catering industry and food a favour, become a plumber you dumb-arse".

Necessity is the Mother of Intervention of an Invention
Steve West

In large establishments and hotels there is a 24-hour ant hill of operators working day and night. A kitchen in one of these fine places was in the basement, not designed by cravat designer stubble thank god, and the restaurant was on the next floor up, street level. Luckily there were 2 small food lifts to transport the food upstairs, yes, every chef living the dream moments as Christmas looms on the horizon, expediting the wares to the restaurant level. It was all very well having lifts, the best we had to work with, but a problem was communication. One system would have been an electronic check machine, that is if one, human error didn't come into play, didn't break down, paper run out.

This, in tandem with a telephone or intercom ensuring quick flow of service, is the order of the day.

Language in our diverse world is wonderful, colourful yet on occasions can be a little, how can one say, disruptive, a Cockney chef, Portuguese waiting staff, Thai still room lady and a German Manager and one world war 2 radio communicator in the mix, really what could go wrong?

Saturday nights with 120, covers, food and checks coming in.

"beep, ZZZZZZZ, YES?

Wahn zzz, bed…over. What the hell.

Juan wants what?

Zzztttt whan zzzttt BED! OBER!

AAHHH, pans clanking, sauces on, hot, steamy and can't understand Captain bloody Kirk."

A succession of running steps indicates someone is coming to the kitchen when the manager spouts, "One bread!".

This was a nightmare.

Something had to be done. In the kitchen the phone and intercom becomes the devil itself as the constant ringing and buzzing during a heated service is usually followed by a barrage of expletives and obscenities and anger nearly on the brink of self-destructing as mind blowing insanity builds up and throwing oneself of the stove to get a night off was nearly on the cards as this chaos was held in deep hatred for the trade.

The checks were dispatched to the kitchen in either a cup with an elastic band wrapped around it as the waft of air could have driven the delicate little paper to the bowels of the lift shaft. The other way, during quieter times the waiter would come down 3 flights of stairs huffing and puffing, 'sshheeck on chef'

A new company had taken over this restaurant which was to be franchised to them by this busy London hotel and as funds were low because of the new venture, investment was tight, and money had been taken in to obviously buy new stuff eventually. Obviously, the waiter could not be coming away from the restaurant and this system, which had been around for some time had to change. Still, checks had to be given to chefs.

The restaurant was a beautiful, grand 80-seater in Earls Court, chandeliers adorned the ceiling and time was kind to this British Empire styled establishment. The restaurant wasn't busy at the time of the franchised take over as the covers on a Saturday night only totalled 30 people. The new franchised company started to take over the Food and Beverage and introduce a new menu based on a Bistro Style concept and the menu reflected the Dordogne region of France with great foods. The word was getting out there and with this new exciting menu, word spread more, and after Fay Maschler gave us a raving review, the covers started to build and the restaurant ambience started to give the restaurant exactly the attention it deserved.

The waiter legs were flagging as they clung on to the side rail in a hope of at least not having a heart attack on the way up the stairs, sweating profusely as the hill climb became more intense. Whispered, breathless coughing "checks on" were becoming frequent.

A change had to be made and we were given, at last, a machine. One of the directors came to the meeting and positively spouted about having an invention and how it was going to make us all happy. Doubts crept in; horror started to invade us. The director came up with a third option. Who the hell knew there was third option, we scratched our heads, carrier pigeons were the only one I could think of? What we didn't realise at the time, this new invention and intervention was to be a blight and horror on all humanity levels and thank all the angels above in heaven I have never witnessed anything like this ever again.

The horrific invasion of our lives and civilisation was simply a 2" plastic tube piping which descended from the restaurant still room directly above the kitchen. Sounds simple? The orders were then given to the lady in the still room, the check is inserted into another plastic tube, the size of a packet of love hearts and send hurtling down the device to the kitchen. But! And this is a big but, all good and well but the check had to stop somewhere, alight in a place in the kitchen for the chefs to cook the food. The intervention of the intervention had been thought through and this wasn't a problem according to the inventor.

The end of the journey for this innocent check to get to us can only be described as all the demons on the planet have descended on us and crawled through hells gates to make sure our lives were jaded and jarred with every plastic tube to come through. No warning given, the first idea of a check coming through gave us palpitations in our chest because we never knew when we would get one. The fear of the unknown far outweighed any service. Nerves raw and exposed as the contents of our stomachs nearly emptying out as orifices exposed and opening in unison nerves unshielded as the chefs would literally cover faces, cowering and darting into corners or hiding places to escape what everyone thought was a raid by a gunman as a shot rang out or a massive bang of a nearby car backfiring or even worse, the kitchen oven exploded and the solid top has been released from its cradle.

Anyone entering the kitchen at this point would only see carnage and madness as chefs and kitchen porters consoling each other, panting and breathless the weeping chefs grabbed by arms being helped up from the previous foetus state they were in.

The cause of this ear shattering sweat inducing onslaught was a simple device of a small biscuit tin, nailed to the wall directly under the plastic piped gun barrel of doom. As small as it was, the shot gun crack was truly,

truly a crime on the human soul. In time, a lot of time getting used to this device was a skill and it certainly aided in keeping one's wits on edge, but to the unsuspecting virgin recipient of this ear shattering momentary drama as it induces a split-second hatred of life.

It did come in handy and had its value even though it really wasn't something one got used to, had its benefits. When a new salesman or rep came to visit or sell something you had already or something that wasn't needed always came to visit during a busy service time. There are 2 types of salesmen, ones who get it and those that don't.

One particular salesman visited the kitchen of the arse of Satan during the middle of service and never came back again. It was the start of the lunchtime service and the lady used to gather around 4 checks, bullets, for us in the kitchen. The machine gun assault was about to commence as the first checks were about to be expedited. The chef had been a bit behind and had only a few minutes to finish his mise en place. The salesman had appeared at the kitchen door and was standing within a few inches of the obscenity trying to make small talk with the chef.

'Er, working here. Don't have time could you leave please?' arms darting about. 'Make an appointment come back another time?'

Smiling the salesman was still there. 'Won't take long chef.'

Chef beckoned one of the other chefs to finish his jobs for him, smiled at the salesman and asked him if he would like a coffee. Eyes of the other chefs were darting about as chef had stopped for a bit. Chef looked at his watch and phoned the restaurant for a "zzzzttt coffee please. Thanks."

The salesman thanked the chef, and self-assured the sell was his, the salesman felt confident. Out of earshot the chef explained to Leo, the restaurant manager, about the situation downstairs. Calmly gently placed the phone down, turned his face smiling to the confident salesman and passed his coffee to him. Upstairs in the stillroom the order had been given as a plan had been sanctioned. Warriors in the field gathered around 10 plastic missiles and in one go all 10 missiles were lobbed down the tube. Down in the kitchen the chef gave a knowing nod to the team as the last thing the salesman saw, perplexed even, were 3 chefs strangely placing fingers in their ears a split second before he hurtled the hotplate, 3 fridges and a commis chef where he was found where he lay at the bottom of the stairs leading to the street, crying and weeping as he kept clutching his chest begging for the pain to go away.

    A young chef calmly gave the weeping salesman his suitcase, and I believe his shoes and placed them down beside him. Shocked and horror struck, the salesman was last seen running up Earls Court Road tripping over himself as he tried to put his shoe back on. He had truly been introduced to the bomb.

Recruiting Staff as Freud Enters the Picture
Steve West

When building a team from scratch in any field of the catering sector, certain skills are needed.

1. Language skills (not foreign); our own mother tongue spread across our own land.

2. Rose tinted glasses; some sights can be pretty unpleasant.

4. Counselling skills and techniques.

5. Broad shoulders.

6. Definitely a huge sense of humour, on occasions, a graveyard.

As a general rule and a quick telling sign, try not to employ anyone who may have knuckles which scrape the floor, they might be really good at cleaning those hard to get at places, but completely useless for anything else, though there is a place for everyone in the industry.

For one large catering football restaurant, 350-seater, we had only two weeks to fill about 15 positions of various levels of kitchen staff to become fully operational before the opening. As one could imagine, there wasn't much time for comprehensive training or completing induction. So in a speed dating fashion, interviews were organised at the local job centre and the prospective applicants started to roll in clutching their pre-requisite CVs. It was wondered if the throng in the waiting room came through the wrong door, when in actual fact these people mistook this as an audition for the living hybrid film for "The Living Dead" and "Little Britain", and the lead part is a one armed kitchen porter named 'Arry.

At this point you may be wondering if you would ever eat in another restaurant again. You'd never eat there again because it's weird people.

The alternative could be to have a dinner party at one's house.

The throng all suited and booted and dressed up to their best fashion-conscience ability, a couture parade of demob suits and knitted ties for the old kitchen and 1970's throwback chefs. More hair than the average barber shop floor.

Students, wearing ripped tee shirts with homemade logos, and the question of "where did all that money invested in education for yourself go, son"? The first telling sign on a particular student and their ability came into question, when asked of his logo, "Down with the W.O.T."! The reply was a surprise, as if the interviewer should have known this, "well, it is against the tyrannical oppression and stupidity of the World Trade Organisation who embraces profits of billions before the rights of humanity causing the distress and plight of..." Right ok... next! The other reason one couldn't believe his credibility... his tee shirt was ironed and clean.

The student, undeterred went on. "...and marches against the large global corporatism like Nike and Levi's profiteering using slave labour, to produce jeans and clothing while the company directors wallow like pigs in shit in bags of obscene money and...".

The drone went on, but the interviewer really did not have time for it.

The chef interviewing didn't have the heart to let him into a little secret. While student do-gooders are standing outside the offices of large companies, with placards and banners, looking down on those wholesome and idealistic young people are the suits, like birdwatchers, binoculars in hand scouring the crowd, pointing, "over there... there, that ugly muppet with the WOT tee shirt", Creative Director shouting to his sketch artist, while hurriedly sketching their next line of "dragged through the hedge backward" look for the following fashion statement of the season.

When asked why the student needed this job, as the restaurant in question didn't demand much skill, just willing hands and a certain amount of common sense, his reply was, "well I need the money… dun I". It would have seemed rude at this stage to cry with laughter out loud, so the stock phrase "don't call us and we will be in touch" was the answer, which actually sub texted means, you're having a laugh, now piss off and don't waste my time.

The interviewer didn't have the heart to let his verbal "wat" drone before him, the acronym was wrong, in fact it missed another 'T'.

After a plethora of applicants, the chef at the job centre was starting to fade, having one more interview to go. The speed dating just needed one more person to fill a final position. The last of the living, walking dammed, strolled through the door and yes, what a surprise, it was yet another student, who happened to be studying psychology. Bearing in mind, on first seeing this person, we noticed his knuckles and arms were in proportion to his body and he seemed bright enough. This had all the indications of at least being on track to making him employable.

The snoozy heat was enveloping the interviewer. After a while, the chef dragged his sleepy head up from his arms, highly disappointed to leave the dream of running naked through a summery, rosemary scented meadow, skipping through icy streams on a hot day with a promise from a huge bosomed woman in a skimpy two piece - kept it clean for the younger readers -, saliva streaming down his face and the marks of his watch against his cheek, Freud was still gabbling on... and on... and on.

The opening of the restaurant was getting nearer, and during the two weeks prior, 'Freud' did come in handy with his counselling techniques on staff, by this stage nearing the opening of the restaurant, we were all getting anxious and nervous.

On one occasion, a chef who was due to start on a particular morning, turned up with his whites and knives in hand. Preparation started: everyone, including the new chef, had tasks and jobs to do and the kitchen was industriously working away. After an hour, the Head Chef came into the kitchen after yet another boring management meeting, to introduce the new General Manager who was there to "turn things around".

One manager replacing yet another, the last one who had the personality and demeanour of an obsessed train spotter with body odour, a fat obnoxious man, who thought he knew it all, when helping out with manual jobs, would sweat horribly, so much so, the arse of his trousers would have a logo of a wet V disappearing into the deep and unknown, the problem was... we knew where.

His management style went under the heading, crisis and how to deal with it, and was as visionary as a blind dog searching for food that isn't there in a room with the light off.

He wrote in his diary when stuff did happen and didn't quite grasp the concept of using the diary in the intention it was created for. Monkey management style, that is, the monkey is the problem. A manager has a problem, the monkey. He doesn't want to deal with it. Ideal worlds would suggest it is because you have the skills and expertise to enable this monkey into something viable, positive and productive. For instance, as a chef or manager in a restaurant has a certain amount of expertise, experience and knowledge to enable the free-flowing customer experience because it is intrinsic to your nature. You know your stuff. An overarching Manager of yours has a problem, could be from head office, middle management of a large company or brand and has been asked for a problem to be solved.

As a view, wouldn't it be prudent to employ someone at that level with enough know-how to get the job done, facilitate the heads of departments below and take on decisions to 'drive' the business forward? One would think, eh. The monkey on this occasion is generally because the higher manager has not got a clue and with the words, 'can you help me with a problem?' as the word 'problem' now becomes your monkey. Of course, you have enough time to save his arse, go above and beyond to ensure his job is safe as your shoulders are driven downwards with the weight of this big humungous baboon, the monkey that is, not the manager.

The new manager, at this venue, however, was totally different in his approach. His extrovert gravelled voice was crusty just like his manner. A strutting Canadian squat all of 5 foot, with inserts in his heels who really believed he was a ladies man, but the ladies thought he was more like a laddie's man.

A consultant, set on by a sports diner where the opening was to take place, the chefs' office in a corner of the kitchen, was in fact a table in the room where all the cables and fuses were.

This hole did have its benefits for the chef, on numerous occasions when the restaurant was pressed to capacity, unknowing to the pebble voice of squat, a fuse would cut out. The best time for this was when 400 standing football fans were watching their team, in the semi-finals of the European Championship, one nil down, Pierce crosses the ball into the middle, Steve Stone hungry for the ball heading his way and... the sound of a Nuclear power station being shut down followed by darkness usually followed by a deep booming echo of angry fans gunning for the manager, much like a scene from Frankenstein's monster, complete with torches and mallets.

Gravel mouth came bolting into the kitchen, and headed straight for the hole in the wall. The main switch mysteriously had cut out. Squat, being the size he was, could only just reach the large switch, so poked his head round the door shouting, in his Canadian Rottweiler growl "HEY YOU"! In the direction of the chef, who just happened to be innocently busy in the kitchen, "DOES THIS BOX AND CABINET ALWAYS HUM LIKE THIS!!?". Chef calmly looked over and responded, "not all the time, but if you're lucky I believe it tap dances to Frank Sinatra".

Blue touch paper lit, Rottweiler bolted back to the junction box, noise of the bloodthirsty throng in the background as he runs, and out of the blue a large stainless-steel flat fell onto the floor, just outside the hole in the wall. Now, toppling onto a quarry tile floor can induce a heart wrenching brain swelling, bottom opening, nausea, similar to driving a car and finding out when the brakes are pressed and all one can see is one's life pass by in a flash.

The extra value of this particular moment came when Squat came out of the room, clutching his chest, gripping his heart to slow it down. His head and torso held against the frame of the door and knees buckling under him, exploding obscenities, which to be honest nobody could understand, but we sort of got the gist of the message. How was anyone to know that just at the moment of the cymbals' crashing, the little Squat had pulled the massive main switch to the "on" position? As he gathered himself and just short of crawling through the kitchen, his suspicious little eyes had nothing to say... for once that was.

The chef looked around at the industrious kitchen and proceeded to walk to the hole in the wall but noticed that there was a person missing. Puzzled, he asked after his whereabouts, and as quick as a flash, out Freud sped up to the chef. Looking slightly shady and guilty, Freud mentioned that Mr AWOL had decided this place wasn't for him. The chef pondered for a moment, "why"? he asked. On hearing the word "why", Freud kicked in. Micro seconds had passed and in his labyrinth mind and supposed wisdom, mentally flicking through past text books of The Naked Ape, John Mills and great philosophers of time gone by, cross referencing extracts from the Beano and Dandy, Freud spouted... "well, chef, in my opinion..." (oh oh!) The bullshitomonitor needle started to wobble slightly.

Freud carried on, "it seemed he had a deep-rooted sense of anguish and under achievement brought on by his step father when he was a child resulting in a relationship that never reached the required expectations of..." Blah blah blah, drone drone... Chef was standing with his hands on his hips, one eye raised.

The whining in the chefs' head got louder, numbing his senses.

"Uh hu". Now, the chef being a caring soul was trying to listen, and he wanted to help and understand his new brigade of children, so he could encourage, develop and build his family into confident young soldiers in the hard battlefield of the kitchen. He tried to listen harder towards the direction of the humming mellow noise as the mouth of Freud was opening and shutting but the chef could hear nothing. Success he could hear at last... "Then... obviously... this anymore.. .chef, CHEF! Are you ok"?

"Right... ok er thanks", the bewildered chef spluttered. Did he actually fall asleep with his eyes open?

Chef was in his office having tea and biscuits, pleased he had solved this difficult and complex situation, although why Freud departed with a smile was still a mystery.

It wasn't until later that chef found out Freud had something to do with the leaving of AWOL, apparently having his own consultations in the changing rooms, probably complete with a couch, notebook and pen in hand.

During the course of the next few weeks, other staff from the kitchen would go missing from time to time, too frequently to be out of the kitchen to go for a piss or a sneaky cigarette. Coincidently, so would Freud. The head chef, being so busy, really didn't notice. The proverbial penny was definitely toppling.

One quiet morning, which in a busy kitchen is rare, the chef decided to take a wander from his stir crazy mind, where usually this is the best opportunity to find something wrong, to find someone in the nearest vicinity whether they did the wrong thing or had nothing to do with wrong thing, give them the biggest bollocking and tell them to get the wrong thing sorted. This can range from cleaning fridges to filling in the correct temperatures in the temperature control sheets, the art of this being, different pens, colours, changing hands to make sure the sheets and charts do not look like they have been filled in on the same day.

Condensing foods is where sauces and items of food are placed into smaller and cleaner containers and in doing so checking for freshness, if not discarding it, making sure there is enough food for service. From this preparation, lists are made, and ordering is done from this simple task.

This condensing is or should be a habitual ritual and should be treated as such. In some kitchens this can be overlooked and if the Head Chef does not notice and it is too late, an opened plastic container of food not taken care of, will evade the kitchen with a strong culminating whiff of a stinking wet dog doused in sweet bubbling fermenting wine. And yes, the food could actually kill someone as well.

If all the above has been done, and there is nothing to shout at anyone for, the last and always the best is the staff changing room. The staff changing room is a complete anomaly as far as the outside world is concerned, simply because it is in a world of its own. The anomaly of the staff changing room is that it does have a mind of its own. It is the start and the finish of the roller coaster ride for caterers.

    The morning is a reminder of what is about to come and indeed the slowest part of the day as opposed to the end of the day or shift, whereby like firemen about to embark on an extreme emergency, clothes are stripped off quicker than hot soap through Vaseline, clothes and whites strewn around the room like spaghetti pasta thrown into a high speed mixing bowl. The reason why this grotty grotto of hell has a mind of its own is it makes all the mess by itself and nobody else. Dirty oven cloths appear from nowhere, stinking trouser bottoms wander in from the street alone and unattended. Chefs jackets, which one could grow roses on, are usually in competition with a horse with stomach problems.

On top of this, the catering changing room has a unique smell all of its own, unlike any other known to man. First there is a base aroma of body smell, socks, and other articles, which to this day scientists would have a problem of breaking down.

The awe-inspiring, magnificence of the caterers changing room smell is that even though the base odour is there, the room, chameleon like, can change. The changes occur because of different days, i.e. when new whites or kitchen cloths are delivered or taken away, who has just changed and of course the obligatory letting one rip first and so on. But above all, and the ultimate king of all that is not decent there is always... always a person who thinks that the Lynx under arm deodorant, which is sprayed all over, much like a smoke machine in a theatre, people coughing and spluttering everywhere, that they bought from a friend at £2 a Litre actually smells NICE, and attracts women. IT REALLY DOESN'T... GET IT! If the person dousing themselves with this obscenity looked half decent, maybe there could be a slim chance of swapping spit with a whippet, never mind a semi-comatose drunk person with a gram of a brain cell.

Chef, changing room bound and electric paddles for the heart ready on entering the grotto, heard voices as he entered. One of the voices just happened to be Freud's, and went unnoticed as chef pushed through the fog of the room he thought he would listen, a sacrifice given the fact that a blue misty haze of smelly socks, and sniff, what was that, something inexplicable had... oh yes Lynx, had hit him square between his eyes.

In a concerned voice Freud spoke "How do you think this has affected you"?

The sobbing reply. "I can't help this.. .this feeling... yet I know it's irrational".

"But you have to come to terms with it... think it through... accept your fears and anxieties, come on breathe deeply, once more, try to relax, we can overcome this".

Chef became slightly concerned that this person, nearly in tears now, had real problems and he felt he had been too hard on Freud. The chef wondered what the problem could be, gender crisis, a death, perhaps? Could his wife have run away, had an affair and this was the end a relationship? Maybe he was married to monkey manager. He would not normally listen in on things so private, but this was obviously serious. Actually worried, he cocked his ear and listened longer.

Freud continued, "But spiders are only part of nature and let's face it they're probably more scared of you rather than the other way around. So, go on pick this one up, they are quite cute in a way".

The sound in the chef's head wasn't an oncoming train, it was blood starting to boil as the noise ricocheted around inside. The blood was now racing as the pounding in his chest began to become louder, and the red mist was rising as he peered around the door and both bodies froze, wide eyed staring at the chef.

In a very, probably too, calm and personable manner, which is not too good, the chef started... in a controlled whisper as the panicking two tried to get past the chef, who was blocking the door, nose to nose with Freud. "Tell me wise one, how long has this been going on for?!"

Freud could hardly speak from his dry mouth, voice box nailed shut with nerves. "Will you answer me... please"? Sarcasm was just retrieved from the chefs' toolbox of life. Freud knew his mouth was open but could only retrieve a squeak. "Mm only a squeak from someone who could talk so much eh"?

"YOU"! Chef darted his face to the accomplice of this tryst of unfortunate event. The other chef started to shake. "So, you think you are scared of spiders"? The other chef nodded, well either that or was still shaking. Chef started to get louder... "Has it ever occurred to you two stupid clowns that human beings are... cute, and why may you ask"? Freud began to acknowledge the reason why but then realise the Chef about to explode didn't give a damn what he thought. "They are cute because they have money, money which is spent in our salubrious and wondrous restaurant, money which pays our wages. Some of those humans, yes on occasion, can seem like freaks of nature but, we accept them. Spiders on the other hand, are freakish as they may be do not have money believe it or not, and in many ways our lives do not depend on these FUCKIN' HIDEOUS CREATURES"! The two, rabbit eyed in the headlights of this bright outburst saw the face of the chef become fire engine red.

Chef carried on. "There may be a way out of this though". The faces of the duo saw a glimmer of hope. "If you can persuade 200 spiders to have a function in our restaurant which includes a 3-course meal, wine and have a disco afterwards, I will only be too pleased to overlook this little love-in meeting of the stupid and moronic society and even throw in a little bonus your way. As this seems extremely FUCKIN' unlikely I suggest you move your arses into the kitchen and feed OUR two-legged animals with eyes on one side of their head and are A DAMN SIGHT LESS HAIRY"!! The two, fighting over each other, scrambled for the door to get out.

After this incident, the chef investigated amongst the staff and established that everyone, except himself, had been analysed. Freud promised it would never happen again, but he couldn't help it, and the chef knew it.

The demise of Freud was growing nearer.

During a busy service the only thing on the mind of a chef is getting the food from the kitchen to the table. This requires the utmost concentration, focusing totally on the pass or the hotplate. On a particularly busy day, beyond anything before, the chef had to be extra vigilant and be on the ball controlling the service, calling out the checks and expediting food to the waiting staff.

A Restaurant Manager thrown into this situation once, described it as playing chess with Anatoly Karpov with two seconds on the clock.

Kitchen was in full flight, checks coming, food going out in all directions and the momentum was in a good, steady rhythm. The timings were perfect, and the service had been going for two hours, but there was still a way to go, when all of a sudden, a table had been returned. A waiter had pushed the wrong table number on the electronic till for a table away and had not noticed. The food came back and was too cold to be sent back out as they had been giving the food a tour of the restaurant searching for the correct customer.

Chef, barking orders, could not go around to the kitchen and had to try and resolve this. Hot under the collar and temper rising, he exploded, shouting out expletives strung together with the odd word of understanding.

The service could not be stopped, the impetus was diminishing, and timings were being knocked back, trying to catch up as the service domino game had started, and the crashing of pans and metals on stoves started to increase with pressure and mirror the mood.

Freud however, from one corner of the kitchen, noticed the volcano Chef "Gory Ramsbottom" was about to erupt, he knew in his own mind, and only his, he could diffuse the situation. The ultimate challenge had arrived. He sincerely wanted to help, to verbally cuddle the chef into calm and tranquillity. Freud had heard of such moments like these. Special people in special hospitals where an inmate's earth wire had come loose, and sanity had jumped the prison wall of the mind.

Freud stopped what he was doing and slowly started to walk towards the shouting lunatic, who is now a smouldering mass of froth and bulging eyes.

He reached the head chef and the noise level had attained supersonic proportions with spit flying in all directions. Freud gently tapped the chef on the arm. The head of the chef shot round swiftly with seemingly no movement, it was there just staring at Freud.

All the other chefs in the kitchen were still working but keeping a slight sideways glance at the stage, which they knew in their minds was about to explode. At this moment, Freud had noticed something unusual about the scene in front of him. A question had entered his mind. Is it possible that a human being can metamorphose into a Tasmanian type of devil creature who breathes fire? Oh "f@#$ing 'ell", what a really strange colour of red that was. He came this far, so he would chance it.

"Chef?" he said quietly, edging towards the ear attached to the smouldering face. Freud was sure he could feel his own nose hairs singeing from the heat of the coal fire face. Whispering now, Freud continued, "Basically, I think you have a problem with the skilful art of communication which has come about..." Chef was about to interject, but Freud feeling confident, shot his finger up to the nose of the chef. In unison the kitchen staff turned away, just like a crowd watching a boxing match when the final finishing blood curdling blow, drops the opponent to the ground. Quickly, one of the chefs shot around the back of the head chef mouthing the words as loudly as possible a hushed whisper could be, "Nnnooo leave it"! At the same time, in sight of Freud, gesturing a silent waving motion with his flat hand across his neck to cut it out and head beckoning and begging him to stop.

The head chef noticed, moved to the signaller who quickly, sharply looked up, whistled and shuffled back.

Chef tried to carry on the service, another tap bounced off the arm. Without looking at Freud, "You still there"? Freud squeaked quietly: "I can see you're troubled". The kitchen went silent.

"OH OHH"!

Freud only stopped for a breath, but also the kitchen became very, very silent.

Within a microsecond, snot, blood and saliva smeared the small window which looked into the restaurant, the pink canvas of flesh crunched against the pane, trapped by a massive hairy hand. The thud of the pink cheek sponge with an eye crying out for help, as the face was bounced back and forth just as if being toyed by one of Harlem Globetrotters...

The chef shook himself from this wonderful dream. There is a point in life when everyone has come across a situation like this. A crossroad if one likes, of being placed in prison for justifiable manslaughter because the prodding of the verbal stick jibes the sense of all that is sane. The other road being, calmly as best as one can… stab him quietly, maybe no one would notice. Ok maybe there are three options, do not cross this very thin line of an anger and rage and with all one's strength deal with it later.

Freud had realised he had misjudged the scenario. Chef straightened himself up and placed his arm around the shoulder of Freud, and as Freud was trying to explain himself, while of course trying to get more oxygen into his lungs, the chef told him to go away. He was trying to get away but somehow his little legs were not touching the ground. The Restaurant Manager came in and caught him, or, actually, prised the fingers from around the neck was more like it.

Momentum of the service starts to back pedal as the last of the checks dribble away from the kitchen. The other chefs start to clean down and disappear for cigarettes and a promise of a jug of liquid amber Valium as a reward.

Freud was normally shy of cleaning down but on this occasion, he wasn't to be seen at all. The head chef walked around the kitchen, and sure enough, there was Freud cleaning the massive walk-in freezer, which just so happened to be the furthest place from the main kitchen and the chef's office. Chef crooks his finger and beckons over to the little lamb, mint sauce already placed next to the roast potatoes on the plate of this person's life.

The moment was to be savoured by the chef, every action slowed down. The damp skull cap was placed on the table, followed by the undoing of his soiled apron, which he placed on the floor, which later he will discreetly throw into the changing rooms and give someone else a bollocking for not placing it in the linen bag.

On reflection, the chef had pondered about the last few hours. It was a young team, inexperienced and the chef had to look at himself. All the warning signs were there, he supposed, and it was obvious to anyone who had not worked in this hot cage day after day, it wasn't easy. The subtle roller coaster ride of the service full of twists and turns where everything is performed with split second timing. Having experience of this ride, one gets to know the twists and turns and is ever ready to hold on for dear life. On the other hand, if one is not shown and taught how to sit on the ride from hell how can one be expected not fall off now and again.

As it happened, he saw Freud in front of him and wondered and wondered, if Chef felt maybe, just maybe, tbat Freud needed a map to guide him, so was it his fault? Nope… so chef sacked him.

Freud wasn't heard of for a while. Then the chef met a mutual acquaintance and asked about his whereabouts and how was he. He was informed that Freud had pursued his degree and eventually passed with honours. Obviously, fair play, he had found his niche in life, fulfilling his dream of helping people out there and dragging them out of the abyss of depression, throwing a lifeline of sanity and hope into this world of uncertainty and woe. Chef was pleased he had followed his heart and hoped he would truly achieve his goals and ambitions.

"So, what is he doing? Where is he?" Asked the chef imagining Freud would be in white lab coat, clipboard in his hand gently oozing people back in the world with his words, his gentleness and kindness in the constant plight of people with mental health issues.

The reply?

"Oh, he sells computers and software to restaurants... apparently he earns more money".

Negotiating the Kitchen Without Killing Oneself
Steve West

When first entering a kitchen, it will take some time to get familiar with the surroundings. Not only for health and safety reasons but also during a busy time, life, limbs, whites and generally keeping one's body parts intact after a working day can seem a very big, painful hill to climb, and one can be victim to the most innocent of objects. As a general rule:

Never cut food on a stainless-steel surface... use a chopping board.

Try to use a knife to cut food.

Try not to cut yourself, for the time at the hotel where I was an apprentice, I thought the butchery department was run by a nurse called Miss Simms, and blue plasters were part of the uniform.

Duck at the right time.

Ducking at the right time is paramount and should not be underestimated, because not all kitchens are built to the size of football pitches. Also, ducking is a life skill for when the chef is standing behind you and for no reason at all will have the urge to slap you around the back of the head, because he just felt like it. The only tiny indicator that something has happened and the swiftness of this hobby, is when a sharp pain has descended into your body, and as your head has realised what has happened, this pain is amazingly followed by a supersonic boom as the chef walks away laughing.

Claiming another notched victim on the chef's office door.

Restaurants are becoming more and more popular with modern themes, new contemporary, eccentric fashions and ideas which are being thought and conceived by owners who decided that the more outrageous and unique the Restaurant, the more people will haemorrhage through the door to see this new haven of eatery heaven. All this however brings in its own demons, and over the years this has brought about the birth of the Designer/consultant, Refurb Guru, Aesthetic Technician Artistic Flouncing Fabric feeling aroma smelling, FFFWWENGGG SHWAYY, EXXPENNSIVE... designers. One would think of course, one would obtain a building, put a few chairs and tables in a room that is nicely decorated and then a kitchen could not be much simpler really... could it?

No, of course not. The concept, luvvie, is to try and be different and become a complete anomaly in the name of style and fashion.

A bank, old houses and old Oxfam shops are in the crosshairs to be changed and transformed into wonderful palaces and bastions of eateries. The plans are being... reborn and put into place by "Cravat Designer Stubble". In this particular story of the "Emperor's new clothes" the owner is excited about how amazing the new designs are, and how privileged one is to be paying top whack for the honour. All very interesting, but there are things to consider.

A restaurant is where people eat, and the menu should be an integral part of the whole experience... one would think. Except these days it is all about maximising turnover, and the Punchline being how much money can be made on each square foot, per hour, per person, per session... per...fection they cry in unison.

On paper the figures are looking very, very good. 400 people x £2.95 x 35 hours a day x 476 days of the year divided by the first number you thought of, x the wife and children's ages combined, double it times the size of Cravat Designers Stubble and his ego... and you are nowhere near the figure you thought you may have. But that does not matter because the owner is ecstatic and happy, that is until he receives his bill for the whole..

.experience, leaving him enough money to employ a monkey to plate and pot wash, no cleaning chemicals, cold water and one marigold glove to be getting on with the job.

But the dream has to live on. Until a bright spark pipes up, "where is the kitchen going"? "Mmm! Damn good question", replies Cravat Designer Stubble junior. The kitchen always seems to be the last place to be thought about. Now at this point, just because the designers have won the most innovative design award for a triangular squash court, doesn't mean it is a good thing.

And here is a thought, how about asking the chef first. There is only one way to design a kitchen, food simply arrives in one door raw and with a few happening bits and pieces in the middle hey presto! It leaves by another door cooked (well, hopefully that is).

A badly designed kitchen will contort the poor chef into bowel crunching, knee twisting, hip swaying, chicaning through tiny gaps and needs high tension springs implanted into their pelvic bone and like Nureyev on speed in a production of Swan lake.

"What about the basement?" squeaks Cravat Designer Jr. "Nnnooo dear boy, out of the question really, what are you talking about, have you learnt nothing"? What he really is saying, is how he can completely bleed the owner dry, leaving him a shell of his former self, and dehydrated and wizened body in the hot dessert of life for vultures to pick his bones. But hey, the figures show, his accounted tells him, is by the time Mars is the new popular holiday destination where Dave Pierce is throwing out dance tracks on radio one, louts are fighting with and getting arrested by the Martian police in the bars... on Mars, he may have enough to retire and live his life in leisure and wonder what the hell went wrong.

The basement in this case, is of course an extremely loose generic term and it means a cave... just, with solidified dinosaur shit, Roman relics of time gone by, and the odd skeleton thrown in for good measure, then the meaning of basement would be absolutely correct. Just checking through the thesaurus, the more accurate term for these hell holes of afterthought include vault, crypt, subterranean vault.

"Better Still" he continued, "let us scoop out the foundations even more, and go deeper and extend". The pot washing monkey at this stage has walked out and has had enough, shaking his head. The verbal tennis match starts. "It's cramped" cosy! "Too hot!", "great in winter and only use the stoves if you need to, it will save money, "but the chef is 6ft 4 inches tall", as the chef disappears at one end of the kitchen, which tapers to a gap of around 11 inches, and is preparing food as if he was Alice in Wonderland looking through the keyhole. "The stainless steel work top doesn't fit"...yet, as Cravat Designer, phone glued to his ear and clutching a leather bound folder, much like the office suits clutch when at meetings, breaking for lunch, having a Triple Decaff, Latte, extra low fat foam, sweetener, and a touch of cinnamon at Stirberks, and "oh I shouldn't but I will", double Chocolate chip, cream filled, ride home on one massive croissant snowed on by icing sugar by Nige, the short sighted new recruit.

"Yes, correct that stainless steel table we ordered, fantastic, so it can be cut, mauled and mangled into a 6-inch trapezoid... Mwah, you're a dear... byyeee". And the phone beeps off, strangely at the same time as the owners will to live. The mindset of the designer is very much like trying to paint the walls of a house which hasn't been built.

The kitchen has been built, and just for a laugh, is one, not overly sized, steel girder with a sharp 2-inch metal bolt sticking out, and it just so happened to be at the right height of any chef working in the area at the time. Now, one would have thought that hitting this offending piece of blight on humanity once would ensure one would never hit it again, but for some reason, nobody ever learnt.

During a busy service in a Restaurant in Kensington Park Road, if listened for carefully, the ghostly sounds from underground, could be heard the cries of past with "ONN ORRDEERR...BOLLOX...BOLLOX...FFFHHUKKINN ELLL...SHI..." followed by a raving mad dervish dance a thud then silence as the chef was knocked unconscious. The pain induced so much anger, that if the chef was not knocked out, he would find the closest person to him and punch the living daylights out of him. This was to share supposedly, and whatever the chef had everyone else was to get it as well.

    If you have never seen or worked in one of these kitchens, no matter how big or small these kitchens are, the criminal inanimate object is always displayed in the most frequent area used. Anyone with an amoeba size of common would have thought to stay away, but the pain on striking one of those brain magnets, or toe stabbers, or even worse, wedding gear removers, is probably the same as driving a hot stave through one's foot.

Chefs from all over London used to meet up and you could always tell which chefs worked in the kitchen of the Seven Dwarfs. Usually, they looked like horror film extras resembling the most gruesome of monsters with stitches and gashes on their foreheads. Keeping company at one table would-be One-Eyed Pete showing his infected trophies to No Balls Maloney from that day.

However, it doesn't stop there with problematic objects. During a busy service, chefs will use every inch of available surface. Turning here, bending there, they condition themselves to negotiating every obstacle very quickly. Swerving to and fro, they lay a mental trail from the fridge to the section, which is usually furthest away than the designer intended to be where the service should be from. A map has been indelibly printed on their memories.

With stealth like agility, order on and off, he goes to the fridge at the service starter gun. Out of the section, turn left, hips to the right, elbows firmly tucked in like a long-distance marathon walker. Slippery bit, skip, (Claus's carrot peelings), right hand down, corner table top, straight on, mop, skip, open door, hand on steak, cling film off, take steak, about turn, back heel door closed, with steak on the plate, and back we go.

Pure poetry in motion...

Ever watchful however, is the kitchen porter, who it just so happened you really pissed off earlier. He is now ready to claim his coveted title of; "CHEF KILLER OF THE DAY".

As the chef is negotiating his way back, the porter pulls out a stainless-steel table. The one with the razor-sharp corner which rips everyone's apron, just an inch or so and sits back to watch. The chef unbeknownst of this is hurrying back and like a blow from a metal Mike Tyson it rips into his fleshy hip jarring the bone and stunning him into total immobility. As rigor mortis sets in and the brain is ringing with excruciating pain, his mouth gaping squeaking, because all his vocabulary is inexplicably missing, like candle smoke in fog. Eyes bulging and secreting fluid of tears, is totally incapable of reacting the Head Chefs abuse for being too slow... the battle continued...

Computers are the new world, or pain in the…
Steve West

In these hurried times of everyday life, everything is getting faster and faster. Work, social life, family life, and life as we know it. To aid the world on its high-speed course and direction this train ride of life, Technology is almost omnipresent in daily lives. Around the 80s no one would have thought a computer and technology would have taken over the catering world. Orders taken, to be read from a screen then food despatched in robotic style from a window direct into your 4 wheeled living room.

These machines are more common now and not restricted to bright rabbit hutches fast food style establishments as large menu items in restaurants are not that different, as the whirring machine noise has replaced the adrenaline pumping CHECK ON CHEF! Nearly gone are the days of handwritten incomprehensible checks as the new age is upon us saving time, paper and a cost, long term effectively will pay dividends to a business.

From a business side, costs and controls are a priority, with profits and slick communication being a must in our business.

Sounds ideal, doesn't it? What could be simpler than a human placing an order on the electronic pad, sent through the ether and landing gently into the machine where the chef acknowledges this message and cooks the food. Dear customer, when you sit at your table and notice the waiting staff taking your order, it is taken with a check pad, taken to the till and then sent. So, with this slick system in place one could believe this is technology actually gets your food out quicker.

Days have gone by when the bar staff would remember what you ordered in a busy bar, go to the Arkwright till, Ching the drawer open and snap it back promptly to give you your change. Nowadays, 4 pints of larger, 1 pint of Hairy Bullock, 3 pints of Nutty Gran and a G and T. Here the games begin. The card or Key comes out, click, punch in the order, 'ere Mary where's the Nutty Gran'? When you would like to shout over, 'PROBABLY UPSTAIRS SHAVING HER LEGS MATE'! But you really don't want to break the attention of the bar person because the bacteria ridden till pad has probably said it isn't on today. Ironically, with customer service so important, how come the bar staff are half cocked looking away from the customer as they try and take your order a meter away as the DJ is belting out tunes in competition with a 747 Jumbo jet runway?

'SORRY MATE WAS THAT A PINT OF GRANS BOLLOCK?'

The order is in the ether and taken and given to the machine. The machine is now churning out a ticker tape parade of paper in the kitchen which the chef will still have to decipher, assuming they have been trained and watched properly. Brilliant, except if the waiter has forgotten to press 'send': then you will be sitting there a couple of days later with cobwebs gathered on your shoulder, your wife has left you and your skinny dehydrated face is just a shell now.

Of course, this is jesting as it never happens.

If you are lucky though, you might get a large American 25 flavoured ice-cream tower sundae glass which seats 4 people with all the trimmings as the band strikes up happy birthday, sparklers sparkling alongside fireworks and a throng of roller skating waiting staff, shouting and whooping barking out the birthday song when all you wanted was a chicken club sandwich and the server put the wrong table number in.

Menus and restaurants are changing concepts and trying to make experiences more exciting while offers of food are getting wilder and more interesting. The smaller restaurants and rustic places tend to head for rustic robust meals. Others can be fine dining, some family establishments with BOGOFs and fairly priced to suit needs on the pocket. When menus are created in some brands a team of highly trained individuals come together, a think tank to hammer out all the details of what to include on menus, what would work, costs and deliver the best offer one can.

The team come together, and this particular brand was American so other concepts had to be in place.

It was deemed if all else fails, get new uniforms for the waiting staff. Miniskirts for the girls, bicycle shorts for the boys. Then if there was any problem with the food the punters eyes and minds would be kept occupied and distracted. For the male's customers, voluptuous females in the skirts and for the female ones a bobbling camel toe in lycra speeding past with a tray of drinks.

When the concept is in place and menu items decided the information has to be inputted into the main system and that is where the salesman comes in. Back in the day, a menu was written, staff were trained and that was it.

The menu offered is only the tip of the iceberg as the information is data driven. Not only can the orders be sent through but also extra messages because not everyone wants everything do, they? There was considerable banter between a chef and bar staff, a friend of mine. I had a demonstration of this while sitting at the bar. My mate said watch this, he's been winding me up all day and I hate him. Watch the kitchen door he explained. He went to the till and punched in some commands. Thirty seconds later the kitchen door to the restaurant burst open with a thud and the chef stood there snorting fire, like a caged lion who had been prodded by sticks through the bars, bloodshot and hanging eyes scouring the restaurant ready to kill but unable to do anything about the situation because of the throng of customers in the restaurant.

The reason why became apparent as the message read

    Table 4   Covers 3

Big fat bastard in the kitchen – you're a pig dog twat.

No dressing, have a nice day…

Computers never go wrong because they are so reliable.

Pig Ear, Sugary Silk Peach
Steve West

There are different types of chefs, that is, Pastry chefs, bakers, etc. It would be rare for a chef to be specialised in all the different types of cooking there is, and of course there are skilled passionate ones who have the trinity of perfection. Pastry chefs are special in their own way and my dad used to be one. Pastry chefs think differently from other chefs and believe it the nature of exacting recipes and precision of the specialist trade and craft. These are the unsung of the unsung in a large brigade. There seem to be seclusion from the rest of the kitchen, cocooned and tucked away from everyone else.

Chefs in the main kitchen can usually knock something up like bread, bread and butter pudding and the basics, and on occasions can create certain desserts to a degree, but pastry chefs have an unmatched skill.

To watch a pâtissier at work on the highest level is an awesome sight, seeing the chocolate being tempered, cajoled into the finest of sights as the shine and crack of the chocolate is comparable to the extreme heat used in ironworks where the rise and fall of the metal changing temperature is being ever watched in order to produce the finest steel. The pastry chef carefully heats the chocolate and pours it onto his marble canvas cooling skilfully with his palette knife enticing the chocolate into submission to the required format he desires. Patience is a word vastly understated in this context: it really is where science meets art.

The chocolate is ready and like a rough diamond ready to be cut, poured into unwashed moulds and when finished, hard as steel and mirror shiny. The sculpture complete, to be admired for a few moments before the chocolate Venus de Milo dies a heroic death on the martyrdom of the dinner table. A true passionate pastry chef has a regal touch. The regal touch is experience on a different level as the hands are his tools knowing the temperatures, knowing timings and the food they work with.

Sugar, the lovely substance, sweet on the palate, added to cornflakes or beverages is taken for granted in the normal world. In a pastry chef's world, there are flowers, baskets and great designs to show off skills and love for what the chefs do.

Anxiety and burden weighed heavily upon the shoulders of the participative audience in this magic show. Concentration from the chef was so powerful on the act about to follow.

The door and hatch were closed shut; the baker's ovens were calibrated as the door of the oven cushioned a leathery large pillow where the fruits of his labour where to finish. The sugar looking and feeling like liquid plasticine ready to be pulled, shoved and blown like glass into anything the master wanted it to be.

The snake of sugary putty is ready to be sculpted: colours added, stretched, influenced by his massive hands, no gloves I might add, stretched again as the arms of the chef outstretched, back in, stretched again as the colours magically start to become real, as the amazing rainbow is being created as if by chance the God Thor himself had been in the kitchen.

The sculpture today is a basket of fruit, a 3-foot-high woven basket complete with handle, containing every fruit imaginable and garnished with shiny, multi coloured rainbow ribbons. It was for a centre stage banquet.

A ball of sugar is placed on the end of a strong giant straw. Chef blows gently turning the shape into a peach. Time and gentleness are the utmost priority as the birth of the peach is imminent and looms into the bright new world. Peach is dusted into icing sugar for the velvety, silky appearance, as the we can all breathe now. The rest of the chef's sideways glance at each other and let our lungs relax.

The process went on for hours. The basket was ready to be adorned with ribbon. It wasn't like Chef had to open a packet and place it on the basket of fruit, no, it had to made and directly placed on to centre piece.

To have been part of this and to see anything like this was an honour and a privilege, a truly wonderful sight. As luck would have it, I was an apprentice, I just happened to be free to be part of the participating audience to help out with the ribbon as the chef needed hands to hold the ribbon while he made up the basket.

The chef in question at the time was a Master, having a gift like no other. A perfectionist, stickler for detail on the highest level and also a disciplinarian, who made everyone wary, except his sous chef who he trusted with his kitchen.

He had no more competitions to compete in, and he had the table of honour at all major hospitality competitions.

With him needing several people to take part of the performance and the hands to hold his precious ribbon, while he deftly turned the sugar into bows breathing, literally, life into an inanimate object, the doors were to remain shut, shutter slammed closed. The operation was about to begin. Everyone had tasks to finish to not detract from the main task in hand.

Everyone had finished except one chef who was free to run around in the cocoon as he was left to deal with dessert orders for the restaurant dispatched through the small hatch. No one was allowed in or out of the toilet sized kitchen. A knock on the hatch was the signal for a check and order for dessert which had to be slipped under the door like a secret love letter. A few minutes later the waiter would return and collect the order which the lone chef could send out quickly shutting the hatch once again.

Situated in one corner of the room was the only telephone which communication was held internally throughout the hotel. Back in the day mobiles were non-existent and occasionally there were phone calls for the other chefs, mates, girlfriends, 'what time do think you'll be ready?' 'Where are you, you're late!' Slam! In those days one could hear phones slamming down and a tone of the phone, a light humming noise giving you the message you are in trouble because you should have been home 3 hours ago. Chefs are never ready on time rotas are really an approximation of what time you will get off work, as one often had to stay on without notice.

As a strict rule personal calls were not accepted. There was a policy of not phoning out or taking calls for any reason. There was another underpinning policy of the staff, if chef wasn't around, it could be chanced, at your own risk though.

'The up against it rules are', 1, you had arranged to go for a drink. 2. Your partner has arranged a night out and is waiting for you, 3. The concert you booked tickets for started 2 hours ago. The feeling of the optimist chef is the internal clock hopes and really believes this task may finish on time and all is right in the world. The realist chef knows it isn't going to happen as these tasks and days like this were ad hoc, you were around, you have hands, still breathing, you're in.

'But Chef', There is point in life when the mouth detaches itself from the brain, separates normality from fantasy. The fantasy in this case was the actual belief you were going to carry on after the 2 words of, "but Chef". The real world has a different idea. On the words, "but chef", an invite has been thrown out to the recipient on his audio letter box of an ear. On hearing this the eyes of the recipient inexplicably just look at you. This look rips into your soul and induces a moment of clarity. That clarity is these two words are only said once in your life. If you are lucky, the chef may speak back, but the ending of this story usually ends with the chef having the final words. But in essence, the silence is usually enough and in the spirit of the moment you submit to your ad hoc event.

Chefs in general start losing friends and mates outside the catering industry because the concept of staying after the time allocated to finish is incomprehensible to them.

Back in the cocoon, every one of the chefs were digging deep hoping the phone would not ring and if it did, everyone hoped it wasn't for them. If it was for you and your name was mentioned, the feeling was one of complete dread and horror. The phone rings, chef is the only one near the phone, he takes it because it could be internal, chef gazes around but never fixed an eye on anyone, 'HE'S BUSY!' phone slammed down nearly knocking it off the wall. The worse thing, not one of us knew who it was for. Each one of us hoping a big hole would swallow us up under our feet and take us now. When you get home, get it in the neck there. 'He was a bit rude, just wanted to speak with you. Can't you tell him where to go?'

A fine chef but one scary bastard. Time was not important to him, never was, as his job was more important.

We were all lined up cradling this ribbon as if our lives depended on it and looking back it probably did. Nobody spoke, quick glances being ever vigilant in not letting our nervous, hysterical laughter bubble to the surface.

The was deathly silence broken by the occasional knock of the secret love letter, dull clanks of utensil against surface as the lone chef continued.

Suddenly the phone rang again, and in the middle of an extremely crucial part of this operation. Once again, the chef was in the vicinity and the only one to answer it. If anyone else was in the area they would have taken the call and whispered a message to carry on to their peers, having their back.

Everyone prayed, glanced at each other and prayed again it wasn't for them. I had this vision of us all legging it like bank thieves escaping, leaving the ribbon in a cartoon mid-air pose as we departed, superhero Flash style, the door, possibly without opening it first.

Chef murmured, 'Mmm, mm, Ah ha, yes, of course dear.' As we all looked like a row of dogs in the stance of cocking our heads to one side, perplexed unable to understand the message.

'Okay, so you want me to pick up a packet of toilet rolls, tin of baked beans and that stuff to remove stains from the bedroom carpet', he muttered in his monotonous nasal tones.

If we had the ability to master the ability to laugh out loud without making a single noise, we absolutely nailed it that night. The basket was finished and displayed in all its beauty and glory.

Us? We went out to Soho for a right good night.

As an apprentice in the pastry kitchen, with all its fine work, detailed workmanship and brilliant people there, as a fresh-faced youth embarking on this new career, an intelligent quota of 2 I felt like Quasimodo modelling swimwear and felt out of place. Wanting to impress and commit to my trade instead of swift, deft and skilful with my hands, it seemed my fingers resembled ten flip flops tied together with chicken wire. At least here I didn't cut myself so much as the butchery.

Yes, another shitty job no one wanted to do. Balls deep in preparing fruit salad, 'did everyone in London only eat fruit salad?' Other important jobs were to watch the flour bins for midges, ever vigil in my commitment, arms folded searching for these damn things. There were always pranks and heard about them before the job became real. 'yeah not going to get caught out.' Laughing inside as the chef instructed me to get 3KG of grapes from the pastry section, but I only need 1KG. 'Ohhkay?' It was instructions which had me in knots on occasions. Walking toward the pastry section, 'He needs only 1K but asked for 3?' Brain in overdrive, I got his 3 and low and behold, he had landed on his section, 1. Chefs never really explained anything.

He spouted, 'Get it!' '

Er not really.

'You see when you got the grapes, brought them back here, how many of those greedy twats nicked the grapes as you walked by them?'

'Er didn't notice chef'.

My skill to bring to the table of the kitchens, naivety, gullibility and not getting stuff. The grapes had to peeled and the seeds taken out. Really? Chef gave me a paper clip and bent it to make the smallest of hooks to take out the seeds. Wow what a brilliant wind up. Such a crazy notion to fall for such a plonker move. Not wanting to get caught out, I didn't do them not wanting to look a right stupid, pointing and laughing objective of the day.

The service started and accompanied the chef on the sauce section, the king of all sections, the premium place to be and prestige and standing of the section and I was there, taking notes, living the real deal in the hotel of all hotels. That was until, chef started darting around his section looking for his peeled and deseeded grapes. As the explosion of the chef subsided, for some inexplicable reason, I found myself in the corner of the kitchen turning around 300KG of vegetables for the service. God bless the 80s and the turned veg era.

Back in the pastry section and a collective of shitty jobs are being carried out, I was always interested in what he was doing and like an irritating buzzing bee around the head, 'What's that for chef, how do you make that?' All so I could transfer the information into my' Adventurous Recipes of the Day' book. Maybe I could try and impress and try these recipes at home, but trying to make 50LB, yes LBs of sweet paste at home wasn't as easy as I thought it could be.

The dialogue generally followed a pattern.

'What's that for chef'?

'Piss off boy'.

'Yes chef, right chef'.

How do you…?

The look came into play. We had an understanding now, on the same level. After a couple of weeks, it clicked, just because he wasn't running around the kitchen like the rest of the chefs, it didn't mean he wasn't busy. Stupidity, naivety and lack of concentration were my strongest points.

One day the chef gave me a compliment, I think. He took me to one side, look straight in my eyes and said, 'Boy. You are the type of person who makes everyone feel better about themselves and frankly I cannot work out whether you are really stupid or really clever.'. 'Thanks Chef'. Rolling his eyes and shaking his head he walked away.

Trying to work out for some time what he meant over a period of time, I did work it out and fooled them, yes, I was actually stupid, I think?

The day and story is coming to an end. Another last shitty job, pot wash for the pastry chef. The equipment pristine, scrapers, mixing bowls, mini car sized, and the sides smattered with chocolate mousse and other goodies for me to index finger food into my gob while the chef wasn't looking. The pot wash was the smallest hand sink as he didn't trust his equipment to the main kitchen porters.

Chocolate moulds were never washed as the tempered chocolate would be released when the time was right.

Now pastry chefs please look away at this moment in time. I learned about not washing the chocolate egg moulds just five moulds in from ensuring the scourer really got all the debris come away from the sides of the metal egg moulds. Needless to say, freeze frame as the sight of the chef hurtling toward me, hands clutched, throttling my neck, Homer Style as my mouth is silently screaming.

The Housekeeper Shouts, Is It a Water Bottle, Is It a Rugby Ball? Oh My God, It's Neither!
Steve West

In hotels, underneath the calm and tranquil world of piped music in the reception and restaurants and the guest bedroom door, there is the promise of a silent, magic world of a pristine room, complete with a swan made towel on your bed inviting serenity and an oasis from a hard day holidaying or conferencing. Mmm, the bath is being filled, the wine is poured and chilled to perfection, you managed to get pass 'Timmy rolly eye' at reception as he was trying to wrestle a giant teddy bear a fiancé had bought for his beloved, but Timmy thought it was a big hairy burglar trying to get away. You slipped by unnoticed. Phew.

On the other side of the customers world is a polar opposite world where it is an unseen battle to ensure the stay goes as smooth and gently as possible. Concierge, bus boys, kitchen porters, chefs, managers, waiters, receptionists and housekeeping, well the list goes on, and all of this only for you, dear customer. You are the only thing that matters in our lives. The answer is always yes, what was the question, dear customer. The customer should never be taken for granted as you give us work, you come into our places of work and hopefully return to give us more work, make us busy because this is what we do.

    Before becoming an apprentice chef, I became a linen porter. My main goal was simply to support the chambermaids in clearing their trolleys of linen and rubbish to put down a chute to the laundry and disposal.. Like it was mentioned, dear customer, you are the most important thing, your welfare was first and foremost. The corridor was a busy hive of activity as the chambermaids maided, the customers, customed and the linen porter, tried to get through the day without hurting himself or indeed others.

The job was brilliant, 'morning' 'Good morning sir, madam. How are you? Can I help you with that?' As like two fleeting ships passing each other. I didn't know you, but my job was to give you attention and respect as if I knew you. That was my first taste of hospitality, and it felt good, really good. I would do anything, try anything as this was a world I loved.

As time went on, I was given more duties and, for a decent amount of overtime, I volunteered for the evening shift. In the hotel there were 8 floors, various types of accommodation, including the master suites, which when first spying these, were a world never been seen with my own eyes.

In the name of customer experience the first priority was to ensure customers were not blown up! Yep, that's right, not hurt in anyway shape of form. At that time there were dangers as it was a high target area. Our job was to keep an eye out for suspicious packages, people and be extra vigilant. The holy grail came to me in one foul swoop, discovering a stash hole of porn magazines, and another object, thinking it was a tennis ball, dragged it out, thinking, which I was told not to do many occasions, and held before me what looked like a child's plastic orange. 'Strange'. 'Looks like an orange with a hole in it'. The orange turned slowly in my hand. 'Weird, as the hole looks like a vagi…F#@%ing' HELLLLL!' As the orange tennis ball shot down the corridor and Beverly the chambermaid picked it up and placed it in the bin. She gave me a knowing look. Not a good one at that.

Stealth and hurried, the corridor walls whizzed past, as my arm filled stash of colourful images, much like Quasimodo limping away from the angry mob at high speed, my cupboard was in sight, the storage cubby hole of delights awaits, at last, turn key, in, back to the door, breathless, safe! My 'currency' collected for a later date. The currency was for food from the main kitchen as the canteen food was not as good as a top club sandwich from Derek in the larder section.

The chambermaids' trolley was spied in the corridor, lined with wallpaper for some reason made of a metallic shiny material. Great fun, not, because enough charge, whilst walking and rubbing souls of your shoes would collect enough charge to blow a small bin through the air and on occasions, if the first 20 times you did this didn't get through my thick skull, I would never get it as walking down the corridor zapping like an exposed 240 watt plug. The linen was overflowing and the big black sack an invitation to be emptied. Linen dumped, back to the rubbish taking care, remember to be vigilant. The large brown bag was seen. 'Shit!' The bag was a size of a filled water bottle, round squashy and taken gingerly from the black sack, next thing radio security. The moment fell silent and with finger and thumb started to peel back the open end of the brown bag. Now, in a situation like that, thoughts racing, danger was foremost in my head.

The bag exposed a pink, oval flesh coloured, well how can this be said, a gentleman's genitalia comforter, so to speak. One end was the insertion point and the other a dial jutted out. The dial offered various settings from gentle hum to, BLOODY HELL! Danger was over but how could I send this down the tube, banging and bouncing down the metal hole.

'Here Beverly, look at this'. Now Beverly was brilliant and such fun and had to be done. Beverly was given the insertion end and the dial was toward me.

'What is that'? 'Er, hot water bottle Beverly'. The dial was ramped up to BLOODY HELL as the comforter started up like a Jumbo Jet engine rumbling down the runway as Beverly ran like a whippet with its tail on fire down the corridor, threw the pink flesh ball in the air, frantically bouncing around expecting The England Rugby team to pick it up and score a try, as the vision of Beverly flailing arms thrashing about shouting, 'MARY MOTHER OF GOD YOU TWAT', echoing down the metal covered walls zapping and cracking away, she couldn't get away quick enough. The comforter was given to lost property, yep the intelligent quota of 2 kicked in. Maintenance guys hung it up in their workshop for a time, until it went missing. Yeah right.

The best thing about being a linen porter was getting tips for cleaning shoes and seeing famous people. Like that time Earth Wind and Fire were leaving the hotel and all the luggage and group were in the corridor, but no one could get past the group. Not because of the amount of luggage they had. It was the hair.

Give Us a Lift
Steve West

Nobody could understand why the chef, who was in charge of the Carvery on the 7th floor of this catering complex, would always ask for one particular young chef to work with him every time. Not that anyone minded, because the chef in question was in fact useless, a weird looking character with an extremely unhuman large appendage also known as a nose. It complimented his face, and as fate would have it, it used most of the flesh on his face and didn't leave a great deal for his deep set eyes, that had a black shadow around all the time. He did indeed look like an evil Gnome.

On top of his facial features, he was short in stature as well, so small that he had to roll up his chefs' trouser legs and jacket three or four times and tie them up with cling film.

With the blight of cling film, which can be every chef's nightmare, as food stuff in containers wrapped so tightly that this caused static electricity. The whizzing of the plastic causing the sharp snap of electricity on anything close to the stainless-steel table zapping around 3000 watts of sparks on the closest part to said table.

Normally average chefs are hip high to preparation tables, but to put into perspective, the table reached the Gnome's belly button.

He wasn't allowed into the carvery to dissect the meat, because as the customers went up to the delicious fare in front perusing the wares and deciding what to eat, the couldn't to be greeted by a talking piece of topside.

However, he was the kind of person that would do anything for anybody. Affable, happy to help and generally a generous chap, but his skills were very much in question. The chef who claimed him, kept him in employment and he was happy with him, so it wasn't questioned.

Most of the main food was prepared, cooked chilled and stored place in a huge main kitchen, which was in the basement of this huge complex. It was so massive it could cater for 2000 3 course meals in one night. The complex had regular chefs and kitchen porters, but because of the size of the establishment, casuals of all levels were employed, almost as much as the regulars because of how busy it was all the time.

There were several floors in this place, the carvery being on the 7th, and another large bistro was on another. Because it was in a theatre complex, the food from the carvery had to be ready for pre-theatre

meals and everything had to be ready for the massive rush after.

One night when one of the chefs didn't turn in for work for the carvery restaurant, I was conscripted, indeed press-ganged to a night on the 7th floor. On walking into the kitchen, the mad scenario made me wonder if I had walked into one of the plays, as a scene from *One flew over the Cuckoo* nest fell before me and questioned my sanity as I witnessed the evening unfold.

    Wondering where the music was coming from, and looking around, the distant blues harmonica seemed to come from behind the kitchen door and the back of the kitchen. Walking in the kitchen, wondering whether to turn quickly around and give my excuses, I entered and was greeted by another chef, nodded and smiled as he kept on working away at his section. He had the sad look of someone who had given up, shoulders drooped, his body language told a visual of a person who has had his soul sucked out of him, and this was as good as it gets in his life. The imaginary aroma of doom invaded me. Poor bastard, I thought.

Knives in hand walking up to the sad-sack, I whispered, 'is that a harmonica I can hear'? He nodded, speech wasn't his forte, and he turned his head in the direction of the music. Slowly edging my head around the corner very carefully, in complete bemusement, turned to the other chef I said, 'is he dangerous'?

I looked again and saw the gnome, legs crossed sitting on top of a work surface soulfully playing, what sounded like 'Home on the range', very badly as well. I returned to the other chef who just shrugged his shoulders, he was sworn to secrecy about the whole thing, but he was a chap who didn't get involved anyway, and he really didn't care. I wondered at this point: was the reason the chef wanted the gnome was to have someone around more stupid than himself. This was the only tangible explanation I could only come up with in this journey of 'the Twilight Zone'.

Not being particularly a busy night, the meals and the pre-show meals had started to calm down around 7pm. The other chef, shruggy sad-sack, and I proceeded to clean down. We were expecting customers later, but not a great deal, and at this point, any help toward me getting to the pub any quicker was appreciated. I tried to ignore the gnome hanging around, his work was a bad as his playing, as he was always very busy trying to find something not to do and look busy in the process of being absolutely lazy.

This is a skill in itself. For anyone who doesn't hasn't had any contact with a person skilful in the art of looking busy when they are not, it is like poetry. It is a matter of placement of said person. The key is to ensure you work with someone who is quicker than you when the Head Chef walks by. This gives a great visual for the Chef as work is being done or looks like it is being done. The other key is, when said task is nearly finished, you should say:, 'okay nearly done, you can finish that', and flit over to another person. And repeat.

The story of the gnome and his service to the industry was becoming clear. Nurse Ratchet, chef was scouring the area like a ticket tout, looking for the police came in and beckoned to the gnome, still darting his eyes around. I was sworn to secrecy, not like put it in a book or anything of the sort. We went over to the service lift which was to transport food through the levels from the basement kitchen and vice-versa.

At the service lift, we gathered as I really was curious now. The size of the lift was 2 by 3 feet, small car boot sized, caged doors which slammed shut on closing. Gnome proceeded to get in, knees tucked up, harmonica still in hand, the chef slammed the cage doors closed.

It was simply this. All but the carvery and the basement kitchen were open. All the other eateries and restaurants were closed and locked after 6 pm. 'Okaaay', starting to back up quietly, still wondering what this was all about, as Ratched started to scare me a bit, as the offering of the day was whispered conversations twinned with slightly bulging, darting eyes. An idea invaded my thoughts. Shall I kill him first before he kills me?

The chef pressed the number 2 on the antiquated lift from the 50s, and off went the gnome. As the lift started to descend, so was the noise of 'home on the range' as it was getting further and fainter.

A distant echo of a dull thud came through the shaft as the faint opening of the cage gates signalled, the gnome must have arrived. Hope he stopped where he wanted, otherwise he could have ended up on the 19.45 from Baker Street.

Yep he arrived.

Followed by the gates opening there was a colossal thud and a distant dulled painful, 'AH BOLLOCKS.' Breaking the brief silent pause was a sound of another thud as the sound of something heavy was put on to the lift.

The lifts doors started to shut downstairs followed by a 'Great Escape' tunnel tapping was heard. The chef pressed his fleshy digit into the number 7 on the button for the lift to return.

Disbelief and amazement were the only two expressions I could hope for at this time as the bizarre scenario unravel before my eyes. Gnome spills out of the lift carrying a case of tinned lager as the face of the chef lights up, really to emphasis this event, he had stolen a squillion pounds worth of stash to end his days in a sun-drenched part of the world surrounded by beauteous maidens.

It occurred the competition of stupidity was in full flow as they all sat there drinking Labatt's, alcohol free beer, cheerily verbally patting each other on the back at this great work they did this evening. The effort put into this was phenomenal, remarkable even, and as I left the kitchen, shaking my head not knowing whether to laugh out loud or cry with pity, the last thing I heard was Gnome and his fucking harmonica playing.

The Grosvenor Hovel
Barrie West

This is dedicated to my dad.

He was a formidable Chef Patisserie, Chef Lecturer and this was a story he wrote so many years ago.

My dad was a crazy one, a 12-yearold in a 60-year-old body, he was the original Peter Pan of humour and personality. Being a prankster himself, his graveyard humour was one of the original 'back in the day' un PC world we don't live in anymore, but nevertheless, it got him through the days when cooking. My dad always said his bosses employed him to keep him off the streets.

These are the musings of my dad, the nutty one, word for word, more or less. Bearing in mind the period was the 60s.

The other night I awoke in a sweat from a nightmare, the extremities of which are hard to describe. I had been transported back over twenty-eight years, to a horror and a hell unknown to what we like to believe as normal, sane people.

It was a prison of London – The Grosvenor Hovel Hotel.

What sins had I committed to wake up screaming in a pool of sweat: to relive the stagnant horrors of captivity and enslavement? Was I to become another George Orwell, writing his descriptions of a Paris Kitchen in his 'Down and out in Paris and London', or Harold Pinter who did his stint at the Hovel, writing his screenplay, 'The Kitchen' and making a bloody fortune to boot.

No, no I could not expose to the sane public the atrocities which, maimed and killed the people who paid to live and eat at the Hovel: people suffering in the name of luxury and decadence.

The premier qualification to work in the dungeons of the Hovel was an intelligence level of 3 and, although there were employees of a higher level of intelligence, it averaged around out at 5.

The hierarchy of the dungeons was such that the more incompetent one was, the better you got on. The worse you got the more difficult it became to replace you. A sort of sideway Murphy's Law.

The activities of the inmates were controlled by Roger Couchie. He was a man of 4feet 6 inches tall and 4 feet 6 inches across. Every morning he would stand by the hotplate at nine and, if one turned up late, he'd scream out,

'Tien's boy, you're bloody late!'

Great way to start the day.

There was the Oberlutentant named Rene Lebeque, but we only saw him twice a year when he wanted a light for his pipe. His headquarters were in a place called the south block, whilst Couchie reigned over the North block which produced food for the first-class pen, also known as The international Sportsmen's Club, and the poor unfortunates in Room Service, Rene looked after banquets. He therefore achieved to hurt people 'en masse' as this was probably the largest facility in London for gathering people together.

Couchie had a number of staff who were laughingly known as sous chefs. Amongst them were One of the sous chefs, a lunatic Scotsman, whose only ambition was to get around all the chambermaids in the hotel, at the last count there were 245, and Willie a mad Irishman totally dedicated to panic, he twitched a great deal.

When we were busy, say fifty cooks in the kitchen and sixteen victims in the restaurant, Willie would run around screaming, "LOOK SHARP BOYS, LOOK SHARP BOYS, WE'RE GOING TO BE BUSY, WE'RE GOING TO BE BUSY". He said everything twice.

The commandant of the restaurant was a tosser called Sonvico. He had been born in the Hovel and rather than talk to a 'cook' he would induce vomiting. The only contact I had with him was when one day he came over to the pastry room, grabbed my arm and said, "Come with me. I'm going to take you onto the restaurant to a table where a lady is sitting. You will doff your bloody cap and bow". He was so precise in his instructions and I was so sure he would go and take a bath having been so close to a cook.

So, he took me to the table with a smarmy, sick smile on his face and I doffed my bloody cap, and I made my bloody bow. The lady gave me a pound note: I could have bloody died. Do you have any idea what a pound meant in those days? 18 packets of Gauloises, 5 gross of French letters, sorry, capote anglaise, 47 fares from Bayswater to Marble Arch on the tube, 27 pints of bitter, 18 fish suppers or 2 hookers in the right part of Paddington. Jesus no wonder Sonvico was pissed.

Anyway, Sonvico threw me back in the kitchen and I found out later the lady had been Mrs Parker, Head of Prudential Assurance. I had made her birthday cake and she deemed it necessary to give the cook a tip. Sonvico never saw fit to mess with me again.

We lived fairly well whilst in this confinement and did our best to eat as much food as possible: also, to steal as much as possible, calling it shopping, to make up for the money we weren't paid. A first commis pâtissier earning 9 pounds 17 shillings and 7d could make his wages up to 27pounds 14 shillings and 6d, and that is without someone punching you out early once in a while.

However, there was always one meal when one was on an early shift that was to be avoided. Breakfast. The breakfast chef was a lonely, miserable person, (as anyone who had to get up at 3am deserves to be), but he used to get his revenge on guests and employees alike. His main advantage was that he couldn't cook. Regardless of how many guests were in the Hovel, and we never knew because no-one ever told us, which as why there was plenty of food around, he would oil half a dozen pans, break 30 eggs onto each one and throw them into the oven. Then he would do the same with the tomato halves and rashers of bacon.

His shift started at 5.30 am and breakfast would be 7.30 'til 10am. All his goodies were cooked by 6am, it doesn't take much imagination to realise the effect on the food. If we wouldn't that for free, imagine the poor bastards paying 3 pounds 9 shillings and 6d to be confronted by a slithery, oil rubbery, hard egg, remember, cooked in the oven on a dirty baking sheet at 5.30 am, plus a dehydrated tomato half and bacon ash. Anyway, he would throw everything away at 10 am and start getting ready for the next day. He was so bad he was almost promoted to sous chef.

He was also responsible for lighting all the ovens in the dungeon but devised a plot to successfully rid himself of this chore. He chose the section furthest away from where he massacred his eggs, which was the roast section, and somehow managed to turn on the gas bypassing the pilots. No sooner had Couchie finished his usual "Tien boys, you're bloody late again", when there was a fantastic explosion, and the six solid tops on the ovens went up into the air like metal frisbees and came crashing down. The oven doors blew off and there were chefs everywhere.

George Thomas, the Roast cook, had lit the pilots and ended up in the sink. Phil, the first commis, was under the table and had crapped himself. Sultan, a four-foot Pakistani, was headfirst in a box of spinach. Couchie was forty feet away hiding in the stillroom. Breakfast cook was in the lavatory on the floor below hoping he had blown up the Hovel.

Needless to say, we either made our own breakfast or got some rolls and coffee from the stillroom: anything to prevent us working before 9.45 am.

The fish corner was an interesting little section with George the fish chef. He was a short man with a temper to match. A bit of a philosopher who would watch a new cook for about six months and, if you were lucky, condescend to say something like "bugger off" if your asked him for something to eat.

His first commis was a guy called George Toplis whose hobby was medicine. Anybody with an infliction either went to, or was sent, to George. George had no social conversation other than medicine, he always spoke in doctors English, even when he was forced to face the fact that he was just a bloody cook. Things like, 'Elevate that Soli Verinquus upon the silver transporter in order that may inspect and elevate its suitability for further traction and eventual consumption by the guest. His way of saying, 'One Sole Veronique'. He was the sort of guy who, if passing through the locker room and coming across a bunch of cooks playing cards would say in a loud voice, 'Gentlemen, please respect your bodies by sitting in a position to avoid contusion of the vertebrae and prevent ending up with lumpus permenantil during your latter years.

When asked about his chef de partie's bad temper, he never tired of explaining about George the fish chef's tumour of the cerebrum and the effect upon the brain. He also had a theory that, after 45 years of messing around with fish and looking at all those dead eyes, it would crack any man.

The last I heard of George Toplis he was teaching professional cookery at the Manchester Hotel School.

The vegetable section was a most interesting section too. It was tightly controlled by Tony and Cisco, (the kid), both being Spanish, and whose motto was to never run out of anything.

Never running out of anything by any normal perspective means having sufficient food to get one through the day, perhaps having a little left over, or even running out and having to cook some more in a hurry. Not so with Tony and Cisco. Their par for the course was a six-week inventory of all foods that could possibly be precooked but, in particular spinach and spaghetti.

They had a standing order with the storekeeper for a daily supply of 15lbs of spaghetti and 5 boxes of fresh spinach and the main priority of the day was to cook these life staples, and really I was utterly convinced if they didn't manage to cook all these products by 11am they would have committed Hari Kari.

I can see it now, either Tony or Cisco, whoever was on the morning shift, would start to worry and demand at 10.30 of the apprentice or commis "Do a we have a enough da spaget, dona runna outa da spaget or Couchie gonna get are arse…de spinch is ready to go, is aqua on for da spinch, make a sure cool it off when cook plis".

Sometimes when I look back, I think this was the only English they spoke.

The cooked spinach was rolled up into elephant sized portions and, with the spaghetti which had been piled onto trays, put into the veg corners cooler which was over by the fish section. To exasperate the situation, Tony and Cisco were not too inclined to walk anywhere. They would delegate this awesome responsibility to an apprentice, or better yet, a kitchen porter who was either, Polish, West Indian or African.

Here lay the problem. Tony and Cisco were Spanish and became problematic in the communication area. It didn't take much imagination.

A Spanish cook telling a Hungarian or West Indian Kitchen porter to put spaget and spinch in el cooler. Maybe it got there, maybe it didn't, it didn't really matter because there was plenty from 6 weeks before. The only problem being that whoever was delegated to put it away, and assuming they got there successfully, would plop it straight on top of the previous supplies.

One night I fancied some spaghetti for dinner and went to the vegetable cooler, opened the door and got a distinct whiff of cabernet sauvignon, 1959 I think, which promptly changed my mind and I went back to the safe boiled eggs.

Tony and Cisco didn't have any problems with Couchie. Once in a while he would complement them. "Tien's Tony and Cisco, vous est un bloody bon job. Vous jamias 86 quelque thing". Well of course they didn't 86 quelque thing because they had plenty of quelque thing, even if it was quietly fermenting.

The vegetable corner was a melting pot of personalities. The staff turnover was high due to a variety of reasons, not withstanding spaghetti and spinach. A new apprentice could be condemned to chopping parsley for six hours. Nobody ever knew where it went or what it was used for, but it had to be done… every day.

The strange thing is if you asked Tony or Cisco for some spinach, spaghetti or chopped parsley, you always got the same reply, "we donna gotta none".

Copper pots were in very short supply, probably because the inventory had never been added to in thirty years, and when anyone left the Hovel, they would take one with them as a souvenir. It was the duty of the night shift to hide the required copper pots for the morning shift. A system which could often break down, particularly if the night shift had spent most of their shift in the staff bar. The pots would be hidden in vents, lockers, ovens, panels in the ceiling and sometimes under the floor.

One morning I didn't have a particular pot – the big one I needed to cook up the next three weeks supply of compote de pommes, which nobody ever ate. So, I told one of the apprentices, nice kid, red hair, to go and find it. If he couldn't find it, make one or go and buy one, but not to come back without the copper pot. He was a conscientious young man, which we soon train him out of, and was gone for a while. The suddenly he came running into the bakeshop, closely followed by Sultan carrying an eighteen-inch butcher knife in his hand. It took six of us to pull him off the apprentice as he was about to plunge the knife into him. He was screaming the only sentence I heard him say. "My copper pot he is taking and killing is what I am doing to him!"

We gave Sultan his pot back and sent the kid to the locker room to recover, but he left, and he got a job pumping gas at a garage in Park Lane. He may be still there. He's the one with the acute twitch.

Tony the vegetable chef had a wife who used to come into the Hovel with him. Nobody ever believed she was on the payroll, but she was there for two square meals a day and to save electric light at home, if they had it, that is. Her main ambition in life was to take the eyes out of the potatoes as they came out of the peeler. This activity was carried out in the vegetable preparation room near the pot wash, or 'plonge' as we aficionados like it to be known. It was well hidden away, a good place for a smoke and a beer. I'm sure Couchie never knew of her presence.

She didn't speak a word of English although she smiled a lot. It was a bit lonely for her so we would talk to her, but she kept on smiling.

One day a ray of sunshine came into her life in the form of Douglas Smith. Douglas had been driven down from Leicester by his parents who had abandoned him on Park Lane because they didn't know what to do with him. Lo and behold, he found himself in the Grosvenor Hovel personnel office and, after the pertinent questions about spaghetti and spinach, Lefevre, Exec Chef, asked him about his hobbies Douglas replied he played guitar, upon which Lefevre hired him as it would be good for morale.

So instead of carrying the usual tools of the trad on his first day, Dougie was carrying a guitar. When challenged by Couchie, "Que ce que vous bloody avez la bas", Douglas relied "Ein Guitar," being really unsure what the Belgian Cockney was saying or what language he was speaking, as Dougie was all the way from Leicester, and it was a long way off. Couchie said, "Bon, c'est pour bloody morale around ici".

It was all uphill for Douglas from then on. There were two types of people as far as he was concerned, those who like him and his music, and all the others. I will always remember the night that Eddie, the Polish aboyeur, the man who calls out the orders to the sane people, called out an order to the sauce section where, unfortunately for Eddie, Dougie was working.

For starters everybody, without exception, hated Eddie. He insisted on calling all the orders out in French although he didn't even master his own mother tongue. He knew everyone hated him, it made him nervous and his voice trembled over the microphone.

"Pppplease chef, faite marcher trois ballotine de vollaile, deux escalope's de veau vienoisse, quattre aloyau de boeuf", and then came to a halt because a six foot 2 inches of Douglas was glaring down at him. "Piss off" Douglas and went to the locker room to play his guitar.

The party never got their food but that was not unusual in the hovel, no one ever seem to complain.

As long as it didn't involve food preparation, Dougie had it made. He would play Fandangos to Mrs Tony and all around the kitchen too. He did like to make fruit salad though: he would dip his hands into the juice to keep his fingers supple.

The Story Never Ends

The spirit and heart of the Hospitality Industry is a strong one. Against all odds people when their backs are against the wall, step up, dig deep to drive themselves for perfection in adverse conditions on occasions, I am proud to be a part of that family, that passion and even after 40 years in the Industry, met great people, cook for great customers, it is a bug which will never leave me.

I sincerely hope you enjoy reading this book as much as I enjoyed reliving great memories.

Wishing you good mental health, peace.

Steve West